Applied Ecology and Human Dimensions in Biological Conservation

Luciano M. Verdade · Maria Carolina Lyra-Jorge
Carlos I. Piña
Editors

Applied Ecology and Human Dimensions in Biological Conservation

Springer

Editors
Luciano M. Verdade
Centro de Energia Nuclear na Agricultura
Universidade de São Paulo
Piracicaba, SP
Brazil

Maria Carolina Lyra-Jorge
Curso de Ciências Biológicas
Universidade de Santo Amaro
São Paulo, SP
Brazil

Carlos I. Piña
Centro de Investigaciones Científicas y Transferencia de Tecnología a la Producción
(CICYTTP-CONICET)/FCyT-UAdER/FCAL-UNER
Entre Ríos
Argentina

ISBN 978-3-642-54750-8 ISBN 978-3-642-54751-5 (eBook)
DOI 10.1007/978-3-642-54751-5
Springer Heidelberg New York Dordrecht London

Library of Congress Control Number: 2014938720

Printed on acid-free paper

Springer is part of Springer Science+Business Media (www.springer.com)

Foreword

The recognition of humans as a major ecological factor modifying the environment can be traced to several decades ago and ecological thinking has gradually developed with the history of conservation efforts. Such framework that focuses on the application of ecological theories, methods, and knowledge to address human-mediated environmental challenges and develop solutions to create a harmonized interaction between people and nature is the main target of Applied Ecology, an umbrella term under which many scientific topics are pursued.

Increasingly, applied ecologists include humans as integral to the systems they study and seek to characterize the relationship between human actions and biological responses, to develop plans to remediate the effects of human actions, or to inform decision-making processes that regulate human activities. However, while among scientists there is awareness that habitats are being fragmented and reduced worldwide, at steady rates and scales, and populations are reducing in numbers and becoming genetically eroded and therefore compromising its long-term persistence, the consequences of this biotic impoverishment to human beings through the loss of biodiversity-based ecosystem services and the consequent decay of entire ecosystems, are being more and more acknowledged by practitioners, decision makers, and society in general. This was recently acknowledged in the world's Strategic Plan for Biodiversity 2011–2020 (Convention of Biological Diversity) where the first strategic goal of the Aichi Biodiversity Targets is to *Address the underlying causes of biodiversity loss by mainstreaming biodiversity across government and society* (http://www.cbd.int/sp/targets/).

In such an era in which biodiversity loss and the ecological consequences of environmental degradation are increasingly unacceptable, integrating natural and social science concerns into conservation still requires additional thought to become more effective in today's human-dominated world. Sustainable and resilient ecosystems need to maintain its ecological structure and function over time while continuing to meet societal needs and expectations. However, ecology and human dimensions have not always been explicitly linked, although each discipline can benefit from the other, and studies related to human dimensions lagged behind research on wildlife populations and habitats and ecosystem functioning.

This book aims at contributing to bridge gaps between the referred disciplines in the context of biological conservation, explicitly incorporating the concepts of both ecological integrity (maintaining the systems structure and function and the species evolutionary potential) and human dimensions (nature society values and user demands). Starting with a chapter where the editors revisit conservation biology concepts and principles and suggest new research directions, the book develops along 13 other chapters in which several contributing authors demonstrate the state of knowledge and illustrate their personal views in three parts: biodiversity-related conceptual approaches (Chaps. 2–6), methodological developments (Chaps. 7–11), and human dimension approaches to decision-making (Chaps. 12–14).

In Part I, the authors address such different themes as the role of history to explain current distribution patterns (using Amazonia as a case study), how sustainable use of resources must account with gene diversity, how species cope with stressors and drivers imposed by changing environments, the role of pathogens and parasites as a part of biodiversity, or how society's agricultural and forestry demands may still translate into farmlands of high natural value. Methods addressed in Part II are also diverse, illustrating either the adaptation of traditional survey methods for application in agroecosystems (bird and medium to large mammals) or review the application of modern technologies (molecular-based tools and stable isotopes analyses) in non-invasive wildlife ecology. Part III addresses human dimensions in the ecological framework, first by incorporating user demands in multi-taxa surveys and, secondly, by focusing on conflict-solving between conservation and the use of biological resources. Decision-making in the conservation context also sets the frame for Part III, where the authors specifically address the need for rigorous population estimates to support resource management decisions and describe a biodiversity-related e-infrastructure that may be a key instrument for national policy development.

Together, they further point out the way to future investigations and identify problems that will need resolution before more progress can be made. This book is useful to wildlife ecologists and managers, facilitates dialogue between science and social scientists, and should support decision-makers.

Margarida Santos-Reis
Centre of Environmental Biology
University of Lisbon
Lisbon
Portugal

Contents

Part I Concepts

1 **Redirections in Conservation Biology** 3
Luciano M. Verdade, Carlos I. Piña and Maria Carolina Lyra-Jorge

2 **Historical Ecology and the Explanation of Diversity: Amazonian Case Studies** 19
William Balée

3 **Phylogenetic Diversity and the Sustainable Use of Biodiversity** 35
Daniel P. Faith and Laura J. Pollock

4 **Adaptation and Evolution in Changing Environments** 53
Luís Miguel Rosalino, Luciano M. Verdade and Maria Carolina Lyra-Jorge

5 **Biodiversity Loss and Infectious Diseases** 73
Kevin D. Lafferty

6 **The Conservation Value of Agricultural Landscapes** 91
Luciano M. Verdade, Marli Penteado, Carla Gheler-Costa, Graziella Dotta, Luís Miguel Rosalino, Vânia Regina Pivello, Carlos I. Piña and Maria Carolina Lyra-Jorge

Part II Innovation

7 **The Use of Molecular Tools in Ecological Studies of Mammalian Carnivores** 105
Francisco Palomares and Begoña Adrados

8 **The Role of Abundance Estimates in Conservation Decision-Making** 117
James D. Nichols

9 **Wildlife Surveys in Agricultural Landscapes: Terrestrial Medium- to Large-Sized Mammals** 133
Maria Carolina Lyra-Jorge, Carla Gheler-Costa, Carlos I. Piña, Luís Miguel Rosalino and Luciano M. Verdade

10 **Point Counts Method for Bird Surveys in Agroecosystems of the State of São Paulo, Southeastern Brazil** 149
Marli Penteado, Wesley R. Silva and Luciano M. Verdade

11 **The Use of Stable Isotopes Analysis in Wildlife Studies** 159
Thiago S. Marques, Neliton R. F. Lara, Plínio B. Camargo, Luciano M. Verdade and Luiz A. Martinelli

Part III Governance

12 **Multi-taxa Surveys: Integrating Ecosystem Processes and User Demands** 177
William E. Magnusson, Ben Lawson, Fabricio Baccaro, Carolina Volkmer de Castilho, J. Guy Castley, Flavia Costa, Debora P. Drucker, Elizabeth Franklin, Albertina P. Lima, Regina Luizão, Fernando Mendonça, Flávia Pezzini, Juliana Schietti, José Julio Toledo, Ana Tourinho, Luciano M. Verdade and Jean-Marc Hero

13 **Who's in Conflict with Whom? Human Dimensions of the Conflicts Involving Wildlife** 189
Silvio Marchini

14 **BIOTA/FAPESP: The Biodiversity Virtual Institute—Translating Research on Biodiversity and Ecosystem Services into Policies in a Megadiverse Country** 211
Carlos A. Joly

Index 225

Part I
Concepts

Chapter 1
Redirections in Conservation Biology

Luciano M. Verdade, Carlos I. Piña and Maria Carolina Lyra-Jorge

Abstract According to Caughley (J Anim Ecol 63:215–244, 1994), there are only four categories of humans' intervention in nature at the population level: biological conservation, control, sustainable use, and monitoring. As the vast majority of the species are not endangered, nor valuable or damaging, monitoring is by far the most relevant of such alternatives. A global network of long-term biodiversity monitoring sites should be established in order to effectively contribute to the decision-making processes concerning biodiversity conservation, use, and control. The following limiting factors should be pursed in terms of conceptual basis: spatial–temporal heterogeneity, human dimensions, adaptation, and the complexity of processes complementarily to the patterns of diversity. In addition, abundance estimates should be improved and the use of molecular markers and stable isotopes should be stimulated to assess ecological and evolutionary processes. Last but not least, governance should be based on the use of populations as units of management and landscapes as units of administration.

L. M. Verdade (✉)
Centro de Energia Nuclear na Agricultura, Universidade de São Paulo,
Caixa Postal 96, Piracicaba, SP 13416-000, Brazil
e-mail: lmverdade@usp.br

C. I. Piña
Centro de Investigaciones Científicas y Transferencia de Tecnología a la Producción, (CICYTTP-CONICET)/FCyT-UAdER/FCAL-UNER, Dr. Materi y España, CP 3105, Diamante, Entre Ríos, Argentina
e-mail: cidcarlos@infoaire.com.ar

M. C. Lyra-Jorge
Curso de Ciências Biológicas, Universidade de Santo Amaro,
Rua Prof. Eneas de Siqueira Neto, 340, São Paulo, SP 04829-300, Brazil
e-mail: carolina_ecologia@hotmail.com

L. M. Verdade et al. (eds.), *Applied Ecology and Human Dimensions in Biological Conservation*, DOI: 10.1007/978-3-642-54751-5_1,

1.1 Introduction

Conservation biology has been considered as a crisis discipline because it deals with the causes and consequences of biodiversity loss (Soulé 1985). In such context, the development of both technological tools and conceptual basis are necessary to perceive, identify, and solve problems. However, how and when should humans intervene in nature is rarely consensual (Hobbs et al. 2011). On the contrary, the debates about such questions often bring conflicting philosophical and scientific views (e.g., naturalism vs. humanism, applied vs. theoretical sciences). However, some points can come up from such debates and effectively contribute to both technological and conceptual development of this field. More recently, Geography and History originated two different perspectives to understand anthropogenic changes in natural environments, respectively, Landscape Ecology (Forman 1995) and Historical Ecology (Balée 2006; Chap. 2 of this volume). Although antagonistic in the way they deal with temporal dimensions, both scientific branches are rather complemental.

Humans' impact on Earth can be compared to the great geological disasters that resulted in mass extinctions (Doughty 2013). However, the comprehension of this as a planetary process is also uniquely human. Such comprehension demands both applied and theoretical scientific development in order to deal with real problems. Intriguing evolutionary questions involve the sometimes surprising adaptive capacity of certain organisms to dwell in altered and/or changing ecosystems that apparently lost most of their structure and functionality (Levins 1968; Ferrière et al. 2004; Chap. 4 of this volume). In such circumstance, not only space but also time in number of generations (Simpson 1944; Gould 1995) should be considered in order to understand the patterns of distribution and abundance of species. It is also necessary to determine at which level (from genes to the landscape, including individuals, populations, communities, and ecosystems) should we intervene in nature in order to identify and solve problems of biodiversity loss.

Notwithstanding, why an extremely successful species as ours, that evolved on a planet where more than 99 % of all species that once existed are already extinct (Sepkoski and Raup 1986), should consider biodiversity loss as a problem? A good reason for such concern would be that never so many species simultaneously lived in this planet (Meffe and Carrol 1994). However, a provocative counterargument would be that, even so, we just superficially know a small part of them (Wilson 1986). Although there is currently a considerable effort to improve our scientific knowledge about the planet's biota, the reasons for such effort transcend the sciences involved (e.g., Ecology, Economy, and others). Such efforts are based on ethical—sometimes actually esthetical—philosophical values eventually antagonistic like naturalism and humanism. However, before having philosophical questions about whether we as a species belong to nature or nature as an abstraction of our intellect belongs to us, we as a species developed many religious views of our origin as such.

1.2 From "Pariahs" to Gardeners of the Garden of Eden

In Western cultures, a platonic view of nature is possibly present at the symbolic Garden of Eden where from our so-to-speak first ancestors have been prophetically expelled for misusing natural resources… Worse than that, after such a biblical fiasco, we had to deal with a demonized wild nature full of horns, fangs, hoofs, and tails… And yet, a human-made ark was the salvation of the wild species in face of a biblical disaster! Such passages and their symbols suggest our long (in human terms) trajectory as an extremely successful species with its religious, philosophical, and scientific paradigms.

Contrary to nature-centered Eastern religions, Western religions are predominantly human-centered (Taylor 2005). In such context, the Western religious view of the relationship between man and nature can be based on three paradigms:

(a) A platonic domestic nature symbolized by the Garden of Eden, created with the single purpose of serving mankind. However, misuse of such resources prevented mankind to remain as its guardian;
(b) An actual wild nature, symbolized by the devil himself with his horns, fangs, hoofs, and tail, apparently created with the single purpose of damming mankind. However, human's history is full of examples of the contrary; and,
(c) In face of biblical disasters, not only human's faith but also humans' skills symbolized by Noah's Ark would be able to save both domestic and wild natures.

These three religious paradigms deeply influenced two modern philosophical views of the relationship between man and nature: humanism and naturalism. The former considers nature as a domesticated part of human *domus*, whereas the later considers mankind as just a part of the wild nature (Hollis 2003). Ironically, the dramatic evolutionary success of humans as a species supports both views. Even more ironically, from an agnostic point of view, one can say we created gods at our own resemblance (Hart 1986) and expelled them from the Garden of Eden. In addition, we fantasized our own wild nature with symbolic horns, fangs, hoofs, and tail (Reventlow and Hoffman 2004). However, we as a species still seem to believe that in face of biblical disasters, only human skills will be able to save both the wild nature and the domestic nature.

Philosophical values have always driven science (Kuhn 1996). As an example, conservation biology has been called a "crisis discipline" (Soulé 1985) in a naturalistic point of view concerning the loss of wilderness currently called *lato sensu* as biodiversity. The analogy of the symbolic Noah's Ark has been used many times in this field with a special concern about the deleterious genetic consequences of populations' small size (Scheuer 1993), possibly the only Noah's mistake…

Such approach has experienced a boom since the late 1970s (e.g., Soulé and Wilcox 1978; Soulé 1986; Fiedler and Jain 1992; Primack 1993; Meffe and Carroll 1994; Caughley and Sinclair 1994; Caughley and Gunn 1995) originating relevant interdisciplinary fields of knowledge such as Conservation Genetics

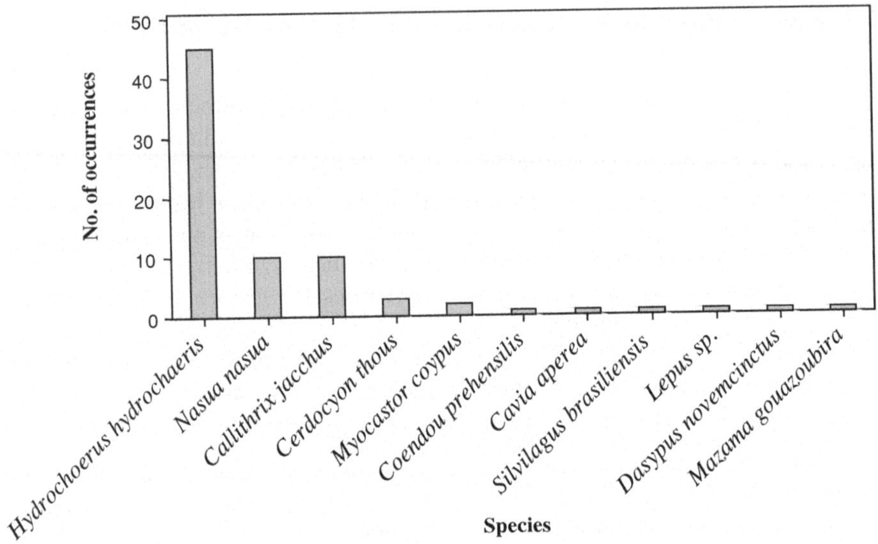

Fig. 1.1 Ordination curve of mammals' on an anthropogenic area of southeastern Brazil (Adapted from Gheler-Costa et al. 2002)

(Schonewald-Cox et al. 1983), Ecological Economics (Costanza 1991), and Conservation Medicine (Aguirre et al. 2002). However, the main paradigm of such approach has been questioned by Graeme Caughley (1994). According to him, in real world, the major drivers of species extinction are demographic (i.e., deterministic) not genetic (i.e., stochastic) processes. As a matter of fact, rarity is common in nature, whereas commonness is rare, as can be seen in many ordination curves, like for example, of middle- to large-sized mammals from an anthropogenic environment of southeastern Brazil (Fig. 1.1). Even in such simple taxocenosis, it is possible to see that three out of 11 species can be considered as common, whereas the other eight can be considered locally rare. The methodological implications of such assumptions are discussed in more detail below (Sect. 1.3.2), but these results are common even for complex communities (May et al. 2007).

The basic differences between the declining population paradigm proposed by Caughley (1994) and the small population paradigm proposed by Soulé (1985) are summarized in Table 1.1. The main limitations recognized by Caughley for the determination of causes of population decline are technological or methodological, whereas for the understanding of how species evolutionarily deal with rarity are theoretical. Not surprisingly, Caughley's approach is usually based on field work and actual data based on hypothesis testing and experimental design, whereas Soulé's approach admits a lot of modeling and meta-analysis. In addition, in order to test hypotheses about population decline, it is conceptually necessary to consider temporal not only spatial heterogeneity (Krebs 1998, 2000).

Possibly, the most important contribution of Caughley to the field of biological conservation is to propose populations—not species—as the units of management.

Table 1.1 Paradigms of biological conservation (Caughley 1994)

	Declining populations	Small populations
Main extinction cause	Demographic	Genetic
Nature of the process	Deterministic	Stochastic
Drivers	Spatial–temporal heterogeneity	Spatial heterogeneity
Unit of management	Population	Landscape
Modus operandi	Field studies	Modeling
Philosophical basis	Humanistic	Naturalistic

Table 1.2 Alternatives of humans' intervention on nature taking populations as the unit of management

Caughley's proposed management actions	Current jargon
Increase depleted populations	Biological conservation
Decrease excessive populations	Control
Establish maximum sustainable yields	Sustainable use
"Nothing but keep an eye on it"	Monitoring

With a few exceptions (e.g., Willians 1996; Dawkins 1989), populations are the unit of the evolutionary process (Mayr 1970). As we discuss below, in order to conserve certain patterns of biological diversity, we should understand the processes that mold them. Evolution is the most relevant of these processes (Mayr 1991).

However, Caughley not only proposed new directions for the area of conservation biology, but he also proposed that this field is only one of the four alternatives of humans' interventions in nature taking populations as the unit of management. As summarized in Table 1.2, such management alternatives would include sustainable use, control, and monitoring.

The unfortunate premature death of Graeme Caughley in 1994 (the same year he published his most impacting publications) prevented the occurrence of a high-level debate in the field of Applied Ecology. Only two years after his death, Caughley has been criticized about his directions in conservation biology (Hedrick et al. 1996). Possibly due to the lack of such debate, the field of biological conservation is still biased to simulation models based on spatial heterogeneity and genetic constraints of small populations instead of on hypothesis testing based on experimental design and field work with collection of real data considering temporal heterogeneity as demographic driver of population decline (Clinchy and Krebs 1997).

The total number of species in the world is estimated in millions (Wilson 1986), whereas the number of endangered, economic, and damaging species is no more than a few thousands (Diamond 2002). That is why monitoring is quite likely the most relevant management action we can take in order to know better—and consequently take better decisions—concerning the species with which we share the planet.

It is noteworthy that Caughley proposed that we should "keep an eye" only on those species that are not endangered, economic, or damaging in case they change their status. However, a misdiagnosis about to what category a certain population belongs (generally, Type II Error) can be disastrous (Magnusson and Mourão 2006).

Considering that most young biologists are not as sharp as Caughley was in counting animals, both in theory and in practice (Caughley 1977), abundance estimates—required to detect population growth (either negative or positive)—most frequently have low precision and unknown accuracy (Abercrombie and Verdade 1995). Even when based on adaptive management, the decision-making process can be ineffective in such circumstance (Magnusson and Mourão 2006).

Public policy concerning biodiversity conservation, control, and sustainable use should also be based on long-term biodiversity monitoring programs (Lindenmayer and Likens 2010). However, it is virtually impossible to monitor everything everywhere all the time (Magurran and McGill 2011). As a consequence, a plethora of indicators have been proposed as surrogates of biodiversity from single species to community level (Lindenmayer and Likens 2010).

From pariahs to gardeners of the Garden of Eden (as suggested by Janzen 1998, 1999), we should keep good eyes on it. In order to do so, we should pursue the limiting factors of biodiversity monitoring. This is quite likely the best we can do in order to improve our interventions in nature.

1.3 Monitoring the Garden of Eden

Biodiversity monitoring should be based on a global network of long-term study sites with an interoperable data bank fed by compatible and comparable datasets (Boyle 2013). In order to do so, either the sampling design should be standardized (e.g., Magnusson et al. 2005; Pezzini et al. 2012; Chap. 12 of this volume) or the estimated individual species abundance should be absolute rather than relative (Chap. 8 of this volume). The main weakness of the former is the eventual difficulty to make researchers change their *modus operandi*. Scientists can be surprisingly conservative! However, the main weakness of the latter are methodological constraints, especially for cryptic species. Such difficulty, on the other hand, demands creativity from scientists. A simple but effective way to do so is occasionally test the assumptions of traditional methods as hypotheses (e.g., Magnusson 1983; Sarkis-Gonçalves et al. 2004).

Like other human activities, biodiversity monitoring can be limited at three levels: conceptual basis, innovation, and governance. The first refers to situations or problems about what we do not know WHAT to do. The second refers to those problems about what we know what to do, but we do not know HOW. The later refers to those kinds of situations we know what and how to do, but we do not know WHO should do it and WHEN. We discuss these limitations in the sections below.

1.3.1 Conceptual Basis

The following conceptual constraints currently limit the implementation of long-term biodiversity monitoring programs.

1.3.1.1 Spatial Heterogeneity Versus Temporal Heterogeneity

The usual lack of historical data about species distribution and abundance usually leads researchers to a simplistic categorization like pristine and anthropic environments. Besides the usual analytical limitations of categorical data (Magnusson 2002a), such approach tends to ignore temporal heterogeneity in both long and short terms (see Chap. 6 of this volume). On the other hand, there is a time lapse among sampling (short term), ecological (midterm), and evolutionary (long term) processes (Preston 1960), which should be considered in monitoring programs. Such concept is particularly relevant in agricultural landscapes where the matrix has a smaller spatial heterogeneity but a higher temporal heterogeneity (Chap. 6 of this volume). Sampling design of biodiversity monitoring programs should be planned in order to detect the sources of variation above. Temporal variation in diversity patterns can be periodical (e.g., seasonal), non-periodical or chaotic (May 1973, 1974), or just based on single events (Taleb 2007). Monitoring programs should be able to identify such patterns (Magnusson et al. 2005; Chap. 12 of this volume).

1.3.1.2 Human Dimensions

When we consider temporal heterogeneity, it is inevitable to consider human dimensions in biodiversity monitoring programs, especially in history, culture, and socioeconomics as drivers of both long-term and short-term land use change. Current patterns of biodiversity abundance and distribution have been deeply influenced by human history even in environments considered pristine such as Neotropical rainforests (Dean 1995; Chap. 2 of this volume) and African savanna (Sinclair 1979), not only in developed regions of the Northern Hemisphere (Chap. 6 of this volume).

In addition, human population growth continues to cause land use change (Lambin and Meyfroidt 2011) for agriculture (Laurance et al. 2014; Verdade et al. 2012), silviculture (Stape et al. 2004), livestock production (Rifkin 1992), and urban development (McKinney 2002). Direct and indirect land use changes are still the main drivers of biodiversity loss (Tscharntke et al. 2005). The human dimensions responsible for them should be considered by biodiversity monitoring programs.

1.3.1.3 Adaptation

Species either adapt or go extinct in face of environmental changes (Chap. 4 of this volume). The adaptive capacity to such changes tends to be species specific (Schluter 2000). However, humans' impact on nature has been increasing in the last millennia becoming planetary in the past century (Doughty 2013). It is, therefore, reasonable to believe that virtually all living species are suffering a strong anthropogenic selection either going extinct or becoming "domesticated"

(Descola 1987). The reality of such process is that human population growth is not only affecting the patterns of distribution and abundance of virtually all species on earth but it is also affecting the evolutionary process that molds such patterns. A philosophical and a scientific question arise from this scenario. The first can be stated as "what do we want to conserve", whereas the second is "how". Do we want to conserve the evolutionary process (Chap. 3 of this volume) or only its results (i.e., the current patterns of species distribution and abundance and their genetic heritage)? If we want to conserve the evolutionary process, we should be able to identify and mitigate the anthropogenic pressures that affect it. Otherwise, we will end up on a planet with only domestic and domesticated species.

Only recently, such concern has been present in the field of biological conservation (Ferrière et al. 2004). However, such concept is paramount for the effective conservation of biological diversity, including parasites and pathogens (Chap. 5 of this volume). Such concept takes into account not only taxonomic diversity but also phylogenetic diversity (see Chap. 3 of this volume). In terms of the policy-making process, the former is biased to specious recent evolutionary groups in so-called hot spots (Meyers et al. 2000), whereas the latter tends to focus on conservative evolutionary lineages.

1.3.1.4 Diversity of Patterns Versus Complexity of Processes

The patterns of biological diversity are the momentary results of the relationship between species composition—and, therefore, species richness—and their relative abundance (Magnusson 2002b; Bonar et al. 2010). Such relationship varies along the time in response to evolutionary processes (discussed above and also in Chaps. 3 and 4 of this volume) and ecological processes, especially in trophic structure (e.g., Verdade et al. 2011) and diseases (Chap. 5 of this volume). The complexity of such processes determines the patterns of biological diversity (Ricklefs and Schluter 1993; Gell-Mann 1994; May et al. 2007). Therefore, we should expect variation in patterns of diversity as normal, not the contrary (Magnusson and Mourão 2006). In order to understand such variation, we should learn how to measure the complexity of the processes that determine them (as discussed below Sect. 1.3.2). Such measurements can be possibly the best surrogates for biodiversity in long-term monitoring programs.

1.3.2 Innovation

The importance of abundance estimates in order to determine diversity patterns and population growth is discussed in Chap. 8 of this volume. The methodological approach should be chosen based on cost-benefit analyses in relation to the questions or goals involved. The estimation of relative abundance indexes is usually simpler than the estimation of absolute abundance (e.g., population density).

However, ecological processes are usually related to absolute—not relative—abundance as they involve biomass and energy flow (Peters 1983, 1991). Therefore, in order to understand (and quantify) the complexity of such process, we should pursue novel technologies and/or methodology that improves our capacity to estimate abundance and biomass. For example, in order to determine trophic structure we should improve our capacity to estimate absolute abundance of predators and preys and also their use of space. Two relatively novel technologies can be used for these purposes: molecular markers (Beeble and Rowe 2004) and stable isotopes (Boecklen et al. 2011).

Molecular markers are usually non-invasive tools that can be used to extract DNA even from scats (Chap. 7 of this volume) in order to identify not only the prey species but also the individual predator (Pompanon et al. 2012). With this information, it is not only possible to determine local trophic structure, but also infer about possible intraspecific variation of it at the predator level. It is also possible to estimate the population size of the predator by an adaptation of a species-incidence curve considering, for instance, the accumulated number of identified individuals in relation to the sampling effort of scats collection. In this case, the estimated local population size would be the asymptote of the model.

The source of carbon and the trophic level at inter- or intraspecific level can be determined by stable isotopes (respectively, C^{13} and N^{15}) (Chap. 11 of this volume). With these information, it might be possible in the near future to identify the origin of the prey even in microgeographic scale in anthropogenic (e.g., agriculture) landscapes where the matrix usually has different C composition than the remnants of native vegetation (Chap. 6 of this volume).

In such circumstance, it would be possible to compare different areas in terms of the complexity of their trophic structure. As an example, Fig. 1.2 shows hypothetical communities before and after human intervention (e.g., land use change). The number and intensity of predator–prey interactions (connective lines in the figure) is a direct measure of the trophic process complexity, which by its turn determines the local pattern of biological diversity. For instance, in the example of Fig. 1.2, a population decline in prey species "b" and "c" can be expected after human intervention.

Molecular markers can also be useful to identify pathogenic species of microorganisms (Ekblom and Galindo 2011). The complexity of the pathogen/parasite–host relationship is analogous to the pattern of predator–prey relationship shown in Fig. 1.2. As suggested in Chap. 5 of this volume, the complexity of such relationship can also be used as a surrogate of biological diversity at the community level.

1.3.3 Governance

A global network of long-term biodiversity monitoring program should be possibly an initiative of the United Nations Environmental Program (UNEP). Signatory countries should establish a common technical protocol and share long-term

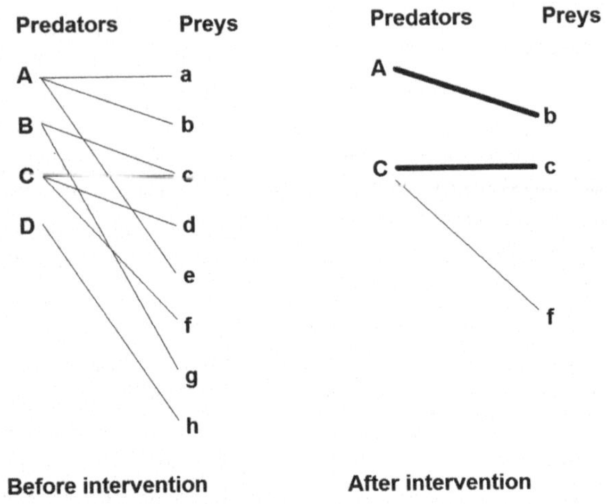

Fig. 1.2 Hypothetical example of trophic structure before and after human intervention (e.g., land use change)

funding for such initiative. Although monitoring programs have the final mission of being incorporated in national and international protocols on environmental governance, a full body of science is required to establish and develop it.

Conceptual and technological constraints have been discussed in previous sections of this chapter (Sects. 1.3.1 and 1.3.2, respectively) and other chapters of this book (Chaps. 12, 13 and 14), including sampling design, data bank, and relevant processes and patterns to be monitored. Last but not least, to be effective in the decision-making process concerning biodiversity conservation, sustainable use, and control, such monitoring program should be operational in two levels, populations and landscapes, connected whenever necessary.

As stated above, as units of evolution, populations should be the units of management at the governance level. On the other hand, as units of land use, landscapes should be the units of administration at the governance level. In order to be effective, the connection between these two levels should include robust indicators, such as area and biomass of native vegetation (see Chap. 6 of this volume) at the landscape level and population/community level (i.e., diversity of patterns and/or complexity of processes). Simple algorithms can be developed for such relationship allowing estimation of both diversity of patterns and complexity of processes based on landscape indicators (e.g., area and biomass of native vegetation remnants).

The administration of landscapes in order to minimize loss of local populations tends to be more cost-effective than the management of all populations separately. On the other hand, discrepancies (e.g., overhunting) could justify specific management actions at population level. Such monitoring program could also be used to guide local and regional policy makers concerning biodiversity and land use (e.g., Joly et al. 2010).

Quoting Graeme Caughley and Daniel Janzen, we should keep an eye on what remains from the Garden of Eden. In order to do so, we should expand our conceptual basis, as well as stimulate innovation and improve governance concerning applied ecology and human dimensions in biological conservation. This is the main goal of this book.

1.4 Final Remarks

Independent of religious beliefs and philosophical values, humans' interventions in nature are ubiquitous. Scientific development should not only expand the limits of conceptual basis and technology, but it should also improve governance concerning biodiversity.

Caughley (1994) suggested we should keep an eye on species that are not endangered, economic, or damaging because they could change their status due to anthropogenic pressures. However, only a few thousand species are endangered and a few hundreds are economic or damaging, whereas the total number of species is estimated in tens of millions on Earth. Therefore, monitoring is quite likely the most relevant management action as it involves the vast majority of the biodiversity.

Biodiversity monitoring should include a global network of long-term study sites based on interoperable datasets and a single data bank, in order to have compatible and comparable data among sites. Such initiative requires the development of the current conceptual basis, innovation, and improvement in local and regional governance.

The following concepts should be improved:

(a) Spatial–temporal heterogeneity should be considered as temporal heterogeneity patterns can be periodical, non-periodical, chaotic, or based on single events, frequently more relevant than the momentary spatial heterogeneity patterns;
(b) Human dimensions such as history, culture, and socioeconomics should be considered as they have been affecting land use for millennia;
(c) Adaptation to anthropogenic pressures should also be considered in a global monitoring program as it affects not only the patterns of species distribution and abundance but also their genetic heritage;
(d) A global monitoring program should be based on the measurement of the complexity of processes instead of the diversity of patterns, as the former determines the later.

The following technological and/or methodological innovation should be pursued:

(a) Molecular markers are usually non-invasive tools to identify species and individuals; therefore, they can be useful to monitor demographic, trophic, behavioral–ecological, and genetic processes at the intra- and inter-population levels;
(b) In order to monitor biological processes, abundance estimates should have higher precision and known accuracy, which requires creativity and inventiveness from researchers;
(c) Stable isotopes (C^{13} and N^{15}) can be useful to determine the sources of C and the trophic level at intra- and inter-population levels.

The following aspects of biodiversity governanceshould be considered:

(a) Populations—not species—should be the unit of management as they are the unit of the evolutionary process;
(b) Landscapes should be the unit of administration as they usually follow distinct public policy concerning land use;

(c) Biomass of native vegetation is possibly the most effective indicator of the complexity of biological processes in pristine environments; and
(d) Area and biomass of the remnants of native vegetation should be tested as indicators of the complexity of biological processes in anthropogenic environments.

Acknowledgements This study was part of the Biota Program of São Paulo Science Foundation (FAPESP, Proc. No. 2066/60954-4). The good ideas here discussed came from delightful talks with the authors of all other chapters. The silly ones are our own.

References

Abercrombie CL, Verdade LM (1995) Dinâmica populacional de crocodilianos: elaboração e uso de modelos. In: Larriera A, Verdade LM (eds) Conservación y Manejo de los Crocodylia de América Latina, vol 1., Fundación Banco BicaSanto Tomé, Santa Fe, pp 33–55
Aguirre AA, Ostfeld RO, Tabor GM, House C, Pearl MC (2002) Conservation medicine: ecological health in practice. Oxford University Press, Oxford
Balée W (2006) The research program of historical ecology. Ann Rev Anthropol 35:75–98. doi:10.1146/annurev.anthro.35.081705.123231
Beeble T, Rowe G (2004) An introduction to molecular ecology. Oxford University Press, Oxford
Boecklen WJ, Yarnes CT, Cook BA, James AC (2011) On the use of stable isotopes in trophic ecology. Annu Rev Ecol Evol Syst 42:411–440. doi:10.1146/annurev-ecolsys-102209-144726
Bonar SA, Fehmi JS, Mercado-Silva N (2010) An overview of sampling issues in species diversity and abundance surveys. In: Magurran AE, McGill BJ (eds) Biological diversity: frontiers in measurement and assessment. Oxford University Press, Oxford, pp 11–24
Boyle J (2013) Biology must develop its own big-data systems. Nature 499:7. doi:10.1038/499007a
Caughley G (1977) Analysis of vertebrate populations. Wiley, Chichester
Caughley G (1994) Directions in conservation biology. J Anim Ecol 63:215–244
Caughley G, Gunn A (1995) Conservation biology in theory and practice. Blackwell Science, Cambridge
Caughley G, Sinclair ARE (1994) Wildlife ecology and management. Blackwell Scientific Publications, Oxford
Clinchy M, Krebs CJ (1997) Viva Caughley! Conserv Biol 11(4):832–883
Costanza R (ed) (1991) Ecological economics: the science and management of sustainability. Columbia University Press, New York
Dawkins R (1989) The selfish gene, 2nd edn. Oxford University Press, New York
Dean W (1995) With broadax and firebrand: the destruction of the Brazilian Atlantic forest. University of California Press, Berkeley
Descola P (1987) La nature domestique: symbolisme et praxis dans l'écologie des Achuar. Editions de la Maison des Sciences de l'Homme. Man New Series 22:754–775
Diamond J (2002) Evolution, consequences and future of plant and animal domestication. Nature 418:700–707. doi:10.1038/nature01019
Doughty CE (2013) Preindustrial human impacts on global and regional environment. Annu Rev Environ Resour 38(2):1–25. doi:10.1146/annurev-environ-032012-095147
Ekblom R, Galindo J (2011) Applications of next generation sequencing in molecular ecology of non-model organisms. Heredity 107:1–15. doi:10.1038/hdy.2010.152
Ferrière R, Dieckmann U, Couvet D (2004) Introduction. In: Ferrière R, Dieckmann U, Couvet D (eds) Evolutionary conservation biology. Cambridge University Press, Cambridge, pp 1–14
Fiedler PL, Jain SK (eds) (1992) Conservation biology: the theory and practice of nature conservation, preservation and management. Chapman and Hall, New York

Forman RTT (1995) Land Mosaics: the ecology of landscapes and regions. Cambridge University Press, Cambridge

Gell-Mann M (1994) The quark and the jaguar: adventures in the simple and the complex. W.H. Freeman, New York

Gheler-Costa C, Verdade LM, Almeida AF (2002) Mamíferos não-voadores do campus "Luiz de Queiroz", Universidade de São Paulo, Piracicaba, Brasil. Rev Bras Zool 19:203–214

Gould SJ (1995) Tempo and mode in the macroevolutionary reconstruction of Darwinism. In: Fitch WM, Ayala J (eds) Tempo and mode in evolution: genetics and paleontology 50 years after Simpson. National Academy Press, Washington, pp 125–144

Hart G (1986) A dictionary of Egyptian gods and goddesses. Routledge & Keagan Paul, London

Hedrick PW, Lacy RCAFW, Soulé ME (1996) Directions in conservation biology: comments on Caughley. Conserv Biol 10(5):1312–1320

Hobbs RJ, Hallett LM, Ehrlich PR, Mooney HA (2011) Intervention ecology: applying ecological science in the twenty-first century. Bioscience 61:442–450. doi:10.1525/bio.2011.61.6.6

Hollis M (2003) Philosophy of social science. In: Bunnin N, Tsui-James EP (eds) The Blackwell companion to philosophy, 2nd edn. Blackwell Publishing, Malden, pp 375–402

Janzen D (1998) Gardenification of wildland nature and the human footprint. Science 279(5355):1312–1313. doi:10.1126/science.279.5355.1312

Janzen D (1999) Gardenification of tropical conserved wildlands: multitasking, multicropping, and multiusers. Proc Natl Acad Sci 96:5987–5994

Joly CA, Rodrigues RR, Metzger JP, Haddad CFB, Verdade LM, Oliveira MC, Bolzani VS (2010) Biodiversity conservation research, training, and policy in São Paulo. Science 328:1358–1359. doi:10.1126/science.1188639

Krebs CJ (1998) Ecological methodology, 2nd edn. Addison Wesley Longman, Menlo Park

Krebs CJ (2000) Hypothesis testing in ecology. In: Boitani L, Fuller TD (eds) Research techniques in animal ecology: controversies and consequences. Columbia University Press, New York, pp 1–14

Kuhn TS (1996 [1962]) The structure of scientific revolutions. The University of Chicago Press. Chicago

Lambin EF, Meyfroidt P (2011) Global land use change, economic globalization, and the looming land scarcity. Proc Natl Acad Sci 108(9):3465–3472. doi:10.1073/pnas.1100480108

Laurance WF, Sayer J, Cassman KG (2014) Agricultural expansion and its impacts on tropical nature. Trends Ecol Evol 29(2):107–116. doi:10.1016/j.tree.2013.12.001

Levins R (1968) Evolution in changing environments. Princeton University Press, Princeton

Lindenmayer DB, Likens GE (2010) Effective ecological monitoring. CSIRO Publishing, Collingwood

Magnusson WE (1983) Size estimates of crocodilians. J Herpetol 17(1):86–88

Magnusson WE (2002a) Categorical ANOVA: strong inference for weak data. Bull Ecolog Soc Am 83:81–86

Magnusson WE (2002b) Community-based ordination: the problem of data handling and interpretation. Bull Ecolog Soc Am 83:77–81

Magnusson WE, Mourão G (2006) Estatística sem matemática, 2nd edn. Editora Planta, Londrina

Magnusson WE, Lima AP, Luizão R, Luizão F, Costa FRC, De Castilho CV, Kinupp VP (2005) RAPELD: a modification of the Gentry method for biodiversity surveys in long-term eco-logical research sites. Biota Neotr 5:19–24

Magurran AE, McGill BJ (2011) Challenges and opportunities in the measurement and assessment of biological diversity. In: Magurran AE, McGill BJ (eds) Biological diversity: frontiers in measurement and assessment. Oxford University Press, Oxford, pp 1–7

May RM (1973) On the relationships among various types of population models. Am Nat 107:46–57

May RM (1974) Biological populations with nonoverlapping generations: stable points, stable cycles and chaos. Science 186:645–647

May RM, Crawley MJ, Sugihara G (2007) Communities: patterns. In: May RM, McLean AR (eds) Theoretical ecology: principles and applications. Oxford University Press, Oxford, pp 111–131

Mayr E (1970) Population, species and evolution. Harvard University Press, Cambridge
Mayr E (1991) One long argument: Charles Darwin and the genesis of modern evolutionary thought. Harvard University Press, Cambridge
McKinney ML (2002) Urbanization, biodiversity and conservation. Bioscience 52(10):883–890. doi:10.1641/0006-3568(2002)052[0883:UBAC]2.0.CO;2
Meffe GK, Carrol CR (1994) Principles of conservation biology. Sinauer, Sunderland
Meyers N, Mittermeier RA, Mittermeier CG, Fonseca GAB, Kent J (2000) Biodiversity hotspots for conservation priorities. Nature 403:853–858. doi:10.1038/35002501
Peters RH (1983) The ecological implications of body size. Cambridge University Press, Cambridge
Peters RH (1991) A critique for ecology. Cambridge University Press, Cambridge
Pezzini FF, Melo PHA, Oliveira DMS, Amorim RX, Figueiredo FOG, Drucker DP, Rodrigues FRO, Zuquim G, Sousa TEL, Costa FRC, Magnusson WE, Sampaio AF, Lima AP, Garcia ARM, Manzatto AG, Nogueira A, Costa CP, Barbosa CEA, Castilho CBCV, Cunha CN, Freitas CG, Cavalcante CO, Brandão DO, Rodrigues DJ, Santos ECPR, Baccaro FB, Ishida FY, Carvalho FA, Moulatlet GM, Guillaumet J-LB, Pinto JLPV, Schietti J, Vale JD, Belger L, Verdade LM, Pansonato MP, Nascimento MT, Santos MCV, Cunha MS, Arruda R, Barbosa RI, Romero RL, Pansini S, Pimentel TP (2012) The Brazilian program for biodiversity research (PPBio) information system. Biodiv Ecol 4:265–274. doi:10.7809/b-e.00083
Pompanon F, Deagle BE, Symondson WOC, Brown DS, Jarman SN, Taberlet P (2012) Who is eating what: diet assessment using next: generation sequencing. Mol Ecol 21:1931–1950. doi:10.1111/j.1365-294X.2011.05403.x
Preston FW (1960) Time and space and the variation of species. Ecology 41:612–627
Primack RB (1993) Essentials of conservation biology. Sinauer Associates, Sunderland
Reventlow H, Hoffman Y (eds) (2004) The problem of evil and its symbols in Jewish and Christian tradition. T & T Clark International, London
Ricklefs RE, Schulter D (1993) Species diversity: regional historical influences. In: Ricklefs RE, Schulter D (eds) Species diversity in ecological communities: historical and geographical perspectives. He University of Chicago Press, Chicago, pp 350–363
Rifkin J (1992) Beyond beef: the rise and fall of the cattle culture. Dutton, New York
Sarkis-Gonçalves F, Castro AMV, Verdade LM (2004) The influence of weather conditions on caiman night-counts. In: Crocodiles. Proceedings of the 17th working meeting of the Crocodile specialist group. IUCN—The World Conservation Union, Gland, pp 387–393
Scheuer JH (1993) Biodiversity: beyond Noah's arks. Conserv Biol 7(1):206–207
Schluter D (2000) The ecology of adaptive radiation. Oxford University Press, Oxford
Schonewald-Cox CM, Chambers SM, MacBryde B, Thomas WL (eds) (1983) Genetics and conservation: a reference for managing wild animals and plant populations. Benjamin Cummings, London
Sepkoski JJ Jr , Raup DM (1986) Periodicity in marine extinction events. In: Elliot DK (ed) Dynamic of extinctions. Willey, New York, pp 3–36
Simpson GG (1944) Tempo and mode in evolution. Columbia University Press, New York
Sinclair ARE (1979) Dynamics of the Serengeti ecosystem: process and pattern. In: Sinclair ARE, Norton-Griffiths M (eds) Serengeti: dynamics of an ecosystem. The University of Chicago Press, Chicago, pp 1–30
Soulé M (1985) What is conservation biology? Bioscience 35(11):727–734
Soulé M (ed) (1986) Conservation biology: the science of scarcity and diversity. Sinauer Associates, Sunderland
Soulé M, Wilcox BA (eds) (1978) Conservation biology: an evolutionary-ecological perspectives. Sinauer Associates, Sunderland
Stape JL, Binkley D, Ryan MG (2004) Eucalyptus production and the supply, use and efficiency of use of water, light and nitrogen across a geographic gradient in Brazil. For Ecol Manag 193(2004):17–31. doi:10.1016/j.foreco.2004.01.020
Taleb NN (2007) The black swan: the impact of the highly improbable. Random House, New York
Taylor B (2005) The encyclopedia of religion and nature. Continuun, London

Tscharntke T, Klein AM, Kruess A, Steffan-Dewenter I, Thies C (2005) Landscape perspectives on agricultural intensification and biodiversity—ecosystem service management. Ecol Lett 8:857–874. doi:10.1111/j.1461-0248.2005.00782.x

Verdade LM, Rosalino LM, Gheler-Costa C, Pedroso NM, Lyra-Jorge MC (2011) Adaptation of mesocarnivores (Mammalia: Carnivora) to agricultural landscapes of Mediterranean Europe and Southeastern Brazil: a trophic perspective. In: Rosalino LM, Gheler-Costa C (eds) Middle-sized carnivores in agricultural landscapes. Nova Science Publishers, New York, pp 1–38

Verdade LM, Gheler-Costa C, Penteado M, Dotta G (2012) The impacts of sugarcane expansion on wildlife in the state of São Paulo, Brazil. J Sustain Bioener Syst 2:138–144. doi:10.4236/jsbs.2012.24020

Willians GO (1996) Adaptation and natural selection: a critique of some evolutionary thought. Princeton University Press, Princeton

Wilson EO (1986) The current state of biological diversity. In: Wilson EO (ed) Biodiversity. National Academy Press, Washington, DC, pp 3–18

Chapter 2
Historical Ecology and the Explanation of Diversity: Amazonian Case Studies

William Balée

Abstract Historical ecology is a research program concerned with the effects of interactions between humans and the environment. These interactions are understood as forms of landscape transformation. Species diversity is one of the principal foci in the historical–ecological study of landscape transformation. In conservation biology, humans are usually not considered to effect increases in diversity except as consequences of secondary succession and the intermediate disturbance hypothesis, if at all. In the study of Amazonia, evidence suggests that humans not only changed forest composition as a result of extensive agriculture (secondary landscape transformation) but also built environments that supported forests that were otherwise nonexistent before human intervention. Human intervention can also account for the existence of some "forest-dependent" species. In light of past human activities and the ensuing effects of these on Amazonian forests, historical ecology provides a working model of explanation of alpha diversity that is more complete than alternative models, including vicariance biogeography, refuge theory, and environmental gradients, when taken in isolation.

2.1 Introduction

Historical ecology is a research program concerned with the effects of interactions between humans and the environment (Balée 2006). By diversity, I mean biological (or species) diversity in given locales, what some ecologists prefer to call "species richness" (e.g., Barlow et al. 2011; cf. Rosenzweig 1995). I prefer "diversity" to "richness" because of the spatio-temporal phenomena that are accommodated,

W. Balée (✉)
Department of Anthropology, 101 Dinwiddie Hall, Tulane University,
New Orleans, LA 70118, USA
e-mail: wbalee@tulane.edu

L. M. Verdade et al. (eds.), *Applied Ecology and Human Dimensions in Biological Conservation*, DOI: 10.1007/978-3-642-54751-5_2,

however, imperfectly, by notions of alpha and beta diversity (Balée 2010; Erickson and Balée 2006; Rosenzweig 1995; Whitaker 1972). It seems evident that a critical debate both in scientific and political communities concerns how biotic and cultural diversity in tropical rainforests today can be maintained for an authentically globalized world society (Crumley 2001; Hornborg and Crumley 2007). The debate has provoked a number of questions. One is what accounts for that diversity in the first place, and especially, as a subsidiary yet unavoidable question, what role if any have humans as a species performed in it, apart from whatever effects humans as a species are having at the present moment? The issue of the human factor in biological and landscape diversification in what was once, in Western thought, considered to be one of the most undisturbed continental contexts, that is, Amazonia, has led to a burgeoning, international literature since the late 1980s (e.g., Balée 1989; Balée and Erickson 2006; Clement and Junqueira 2010; Denevan 1992, 2001; Erickson 1995, 2000; Heckenberger 2006; McEwan et al. 2001; Pärssinen and Korpisaari 2003; Raffles 2002; Stahl 2002). Popular media have taken up the new view (Mann 2002, 2005; Sington 2002). This varies somewhat from the belief that humans are detrimental to biodiversity and that their effects tend to simplify and even destroy landscapes. To be sure, ever since Marsh (1885) saw humankind as a landscape-maker (in coining the phrase "Man makes the Earth"), the concept in general has not been new. Yet to apply it to a seemingly untouched wilderness, such as Amazonia, is for the history of ideas recent. And to suggest that humans may enhance landscape diversification and speciosity has even seemed to be an even more radical, if misunderstood (e.g., Barlow et al. 2011; Bush et al. 2007; McMichael et al. 2012) claim. How does one answer the question of diversity's origins? The Amazon Basin contains the largest contiguous expanse of tropical rainforest in the world, and within that rainforest, many forests can be discerned in terms of different suites of species, climatic conditions, edaphic structures, and human societies, which do not all have the same impacts on the land and the biota found on and in it. Amazonian diversity is a "riddle" (Bush 1994). Like most riddles, it has no simple answer, but rather a nuanced, multi-causal one.

2.1.1 Society, Time, and Diversity

Directing inquiry into origins of diversity on an Amazonian scale requires a sophisticated understanding of time, that is, time in more than one category. Amazon diversity in its temporal dimension is not just a natural science question and can in fact never be answered fully by the axioms, methods, and techniques of natural science alone. Actually, the so-called social sciences have staked out a crucial territory of data that are needed to explain biological diversity, and by that, I mean the extant number of genotypes in a given region. The assertion about the relevance of social science to biodiversity only seems to be paradoxical because of most scientists' unwillingness or inability to cross disciplinary barriers, and because of a still long-held belief by many natural scientists in a rigid separation between nature and culture and the various iterations of that dichotomy. As a case point, McMichael et al. (2012) suggested that alpha diversity might have been increased here and there by indigenous archeological cultures of western Amazonia, but overall, such an impact was negligible and the diversity of the forest cannot be explained by a human presence in the past (also see Bush et al. 2007). The main problem with this interpretation of the data concerns the baseline: The supposed original forest might itself have been a cultural one, in light of long-term occupation of Amazonia by humans, dating from the Pleistocene. Historical ecologists—at least, I think, most historical ecologists—propose that a barrier between social and natural science is empirically false and tends to represent a mystifying obstacle to a more comprehensive and accurate understanding not only of biological diversity in a global context but also of the range, distribution, and changes in regional cultural forms within a diachronic framework as these forms interact with environmental changes that have been sometimes induced for the long term by human beings, such as arguably occurred in the Amazon Basin (e.g., Balée 2010; Denevan 2001; Erickson and Balée 2006). In other words, time is really multi-dimensional in historical ecology.

Partly to understand time in this complex sense, it seems reasonable to refine first the concept of "humans as a species," regardless of whatever shared nature we have thanks to natural selection and the fortuitous appearance and radiation of anatomically modern humans some one hundred to two hundred thousand years ago. The species falls into an assortment of types of sociopolitical and economic entities, such as egalitarian societies (including many foragers and horticulturalists), ranked societies with weak chiefs exhibiting simple chiefdoms, ranked societies with strong chiefs and complex chiefdoms, state societies composed of tiny priestly elites and vast peasantries dependent on intensive agriculture, and industrial and postindustrial states with social classes, occupational specializations, hierarchies of wealth and ownership, and other inequalities of multitudinous varieties. Instantiations of these and other types of socioeconomic entities can be adduced in the archeological, ethnohistoric, and ethnographic records [on the importance of typology, regardless of whether it is culture-evolutionary, see Earle (2002: 45)]. This differentiation does not mean that they do not overlap or co-exist for certain periods in certain locales.

It is necessary to remember that these are types and not rigid categories. In the Amazon case, it is difficult in a contemporary sense to distinguish historically among the socioeconomic and perhaps ethnic types called "caboclo" and "colonist" (Brondizio and Siqueira 1997), though one can distinguish unlike environmental impacts that have been termed "caboclo" and "colonist" footprints, respectively, since these have been readable by remote sensing technology (Brondizio et al. 2002). Least understood of all such "footprints" are those of indigenous peoples who long preceded the peasantries and urban populations of the Amazon River. However, one looks at history, it cannot be dismissed that Amazon peasantries by definition were always connected to the world capitalist system [they were on its *periphery*, whereas indigenous societies are typically in the *external arena* of the world system (Wallerstein 1974: 332–339)]. The reason the indigenous footprint has been so obscure is probably not because they made no environmental impression at all, which is a doctrine central to the adaptationist model (see Appendix) of Amazonian diversity (Barlow et al. 2011; Bush et al. 2007; Meggers 1996; Moran 2000; McMichael et al. 2012; see Balée 2010; Balée and Erickson 2006), but rather because remote sensing has not yet distinguished between very old forest fallow and primary forest, though I hope it will be able to do so soon. When it does, we will see a much more substantial landscape signature of ancient societies than is today recognized. Ground-truthing has done that up to a point. What proponents of the adaptationist model [i.e., the standard model (Stahl 2002; Viveiros de Castro 1996) or cultural ecology] call secondary succession is really only very recent secondary forest, better classified as old swidden, not as "forest" per se (Balée 1994); it is not cultural or anthropogenic forest in the sense of forest relics left by earlier societies (Balée 1989). The indigenous footprint exists in a longer, more sweeping timeline than caboclos, colonists, and other socioeconomic entities connected to the world system at any point in time since the sixteenth century; even when there have been influences on indigenous systems originating from outside Amazonia, the impacts have been less intensive and less obvious than in the other two cases, until recently (see Fisher 2000).

2.1.2 *Time as a Multidimensional Analytic Phenomenon*

As these types are not rigid categories but heuristic entities for the purpose of differentiation of environmental impacts, exemplars of these types have also had histories and timelines. Varying developmental concepts of time need to be distinguished in order to understand their histories and effects on the landscape. Time in its mythical versus historical and linear versus nonlinear or cyclical senses of time in the emic analysis of the past, which among diverse Amazon cultures, is well presented in Whitehead (2003), and the various chapters in that work lend support one of my main assumptions here, namely, that time is more than one thing categorically. That assumption, in turn, derives from Fernand Braudel's concept of the *longue durée* (1993), involving linear processes that take hundreds and sometimes thousands of years to complete in human history. Braudel explicitly recognized that time, for historical purposes, exists in more than one category. I would further argue that for the purpose of understanding Amazon landscapes and diversity today, the timelines are more numerous, involving in some cases millions of years (for species' genotypes within the school of vicariance biogeography, and tens of thousands or fewer thousands of years within the school of Pleistocene refuge theory). Although that naturalistic time frame for Amazonia is not human historical, it is historical in a broad sense (considering evolution to be a kind of history) and it is relevant to the species distributions first encountered and later modified by human beings thousands of years ago when they first set foot in Amazonia.

Another conceptual usage of time is pertinent here, for it represents a historical moment in Western thought concerning living alterity (contemporary "others") ensconced paradoxically in the past and representing Western origins. McGrane (1989: 104–105) incisively summarized this nineteenth-century usage thus:

> As Lyell's geology, following Linnaeus, was the first to massively and intrinsically include the element of time *within* the very definition of geography (the earth's surface is not primeval but the daughter of time), and as Darwin's biology, following Lamarck, was the first to intrinsically include the element of time immanently within the very definition of the species, so nineteenth-century anthropology was the first to splice the dimension of time immanently inside the experience and definition of the otherness of the non-European Other, inside the definition of traveling beyond Europe, and inside the definition of European "civilization" or "culture".

In the nineteenth century, different geographic areas that included cultural and social Others constituted in and of themselves a time machine, and in this sense, time and space were merged into a single conceptual framework (McGrane 1989), in a pre-Newtonian way.

Clearly, the impacts that dissimilar sociopolitical entities encompassing human populations through history can have on natural environments and on the resources drawn up into the vortex of the world system's demand on raw materials are different. This has been so at least since the emergence of a capitalist world system in the sixteenth century (Wallerstein 1974) together with an emic understanding of Europe (and inner Asia distinct and separate from the rest of the world and therefore not insular land masses surrounded by a tempestuous, threatening, and unknown Ocean (McGrane 1989: 34–35). But the emergence of states in general is probably the critical factor in decreasing biological and other kinds of diversity, at the species level and below it. That is why the study of the interactions of humans and the environment takes on such complexity: Human activities are framed in the context of extremely differentiated social, economic, and political complexity, and sometimes these differences are marked linguistically and culturally. Some impacts may affect landscapes in such manner as to enhance their total number within the environment of a local society and its people and traditions, as well as to even increase species diversity (in terms of the alpha, sometimes beta, but not usually gamma indices), whereas other impacts from dissimilar entities (such as global capitalistic, industrialized, and information-age society) may often have the local effect on tropical forests of diminishing diversity of both landscapes and species (Balée 1998).

But the arguments on origins of Amazonian forests and diversity (and the debate applies with modification to tropical rainforests more generally—for West Africa, see Fairhead and Leach 1996) are not simply drawn up as oppositions between nurture and nature, or culture and biology, or history and evolution (Whitehead 1998). A significant reassessment of the time frame of evolution—both of the landscape and of species—has been underway in the last several years regarding Amazonia. Amazonia has been well into the twentieth century and for many people still is a region of "people without history" (see Wolf 1982 for the original definition). This view is a continuation of the nineteenth-century notion of the non-European. Other, trapped in savagery, frozen in geographic space that really represents "our" own time long past (McGrane 1989; Heckenberger 2006). The reason for the failure by many to recognize historical (and human) process in forming landscapes of Amazonia is because Amazonia in historical comparison with the Andes and Mesoamerica lacked centralized authority. The revisionist school of historical ecology argues that this seeming lack of centralized authority is an interval of history following epidemic disease in the mid to late sixteenth century that essentially wiped out complex society before it could be studied. For that reason, as Whitehead (2003: vii) put it, Amazonia mistakenly still "exists in an eternal present of 'first contacts' and 'marvelous discovery'." But we are beginning to see the time depth and the historical impact of humans on Amazonian landscapes of the past (as in McEwan et al. 2001) (see Appendix).

2.2 Origins of Diversity

For understanding diversity, even biological diversity, essentially the problem hinges on what is an origin, and can or should it be contextualized? Is it a beginning of a genotype regardless of its spatial context, or is it the explanation for why an extant genotype exists in a place with specifiable boundaries? Both understandings of origin—the purely diachronic and the diachronic combined with the diatopic (variation in space)—can be conceptualized more holistically than has heretofore been done in order to grapple with the diversity today of genotypes in Amazonian tropical forests, and indeed, in other tropical forests.

Let me briefly review the findings of the four hypotheses on Amazonian diversity—both of landscapes and species—and then propose what I think will be a more comprehensive alternative, one that is inspired in the research program (see Appendix) of historical ecology which promises not only a fuller understanding of biotic diversity but of cultural and sociological diversity as well. From that point, I think we will begin to see that biology and culture share common fates, and their recent histories so are deeply intertwined that when Amazonia is taken as a regional object of analysis, the two cannot be fruitfully understood apart from the other.

There is an essential dichotomy between the first three hypotheses and the fourth centering on human effects on nature—the dichotomy is simply stated as, for the first three, there was none of significance.

2.2.1 *Natural Kinds of Explanation*

Scientists working within the school of thought called vicariance biogeography have studied speciation concepts in a deep time frame. History and "historical" biogeography (Bates 2001) occur over time periods of millions, not thousands of years. The evidence derives from tectonic events, the Andean orogeny, marine transgressions, and the connection of North America to South America via the uplift of the Isthmus of Panama (Mörner et al. 2001; cf. Lovejoy et al. 1998; Räsäsen et al. 1995). These researchers tend not to rely on fossil evidence, which unfortunately is almost wholly lacking, but rather on data from molecular phylogenetics. Vicariance biogeography is a standing critique of Pleistocene refuge theory, based mostly on the time frame of speciation (Colinvaux et al. 2000; cf. Haffer 2001).

Pleistocene refuge theory derives from repeated observations of endemism in a wide variety of taxa. Problems of speciation have been seen in terms of forest reductions, allowing for genetic drift within remaining patches of forest during colder periods in the Pleistocene. Haffer (1969) originally proposed the hypothesis, and numerous scientists in diverse fields (of entomology, botany, herpetology) soon found patterns of endemism among the groups of taxa in which they specialized, and to some extent, the identified refugia of diverse taxa overlapped. One of the principal early critiques of Pleistocene refuge theory concerned its method: It arguably had a sampling bias (collections were made near major cities, such as Manaus and Belém; hence, refugia are found there, and more seldom in interior, isolated areas—see Nelson et al. 1990). Nevertheless, certain primitive organisms still occur in restricted locales, such as cycads, which appear in all refugia thus far identified but not outside them (Daly and Silveira 2002: 59).

A third long-standing hypothesis on Amazon diversity concerns simply the requirements for tropical moist forest (Whitmore 1990). Environmental gradients, which interact and overlap in producing environments as we know them, include latitude, rainfall, temperature, light, and soils. If latitude alone is used, tropical deserts are low in diversity (Begon et al. 1990: 835) in contrast to tropical moist forests. If rainfall alone is used, high rainfall in tropical Asia actually coincides with low diversity (Gentry 1988). The Chocó in the Neotropics receives as much as 10,000 mm rainfall/year, but it is not appreciably higher in alpha and beta diversities than areas getting around 4,000 mm/year on the eastern side of the Andes (Gentry 1988). None of the gradients permit the restriction of gene flow (Colinvaux 1987) which is needed for allopatry and evolution: drift therefore cannot occur under such conditions. Environmental gradients nevertheless represent necessary if not sufficient conditions for explaining diversity in tropical moist forests. At the alpha scale, sufficient conditions may involve human activity (Balée 1989; Denevan 2001; Erickson 2000; Heckenberger 2006; Heckenberger et al. 2008; McEwan et al. 2001).

2.2.2 *Historical Ecology and Anthropogenic (Cultural) Forest*

What is a forest that is determined by human and cultural activity, and how is it different from any other not so caused? This is a difficult, perhaps tendentious question, but it should be resolved before one can undertake systematic analysis of Amazon diversity and perhaps to diversity of other tropical forests in other parts of the world. Human history in the environment is the factor behind diversity that is least understood of all, but if incorporated into a general model of Amazonian diversity, based on historical ecology, it could bring all current models into mutual understanding. The essential point is humans moved biological diversity around; they also engaged in the domestication, semi-domestication, and cultivation of species—this had the result of transforming landscapes through time (Clement 1999a, b; Balée and Erickson 2006; Erickson 2006, 2008).

We can see that climate change science has well demonstrated current human influence on species' plentitude and on the biosphere generally to be greater than at any time in history. Our geological epoch, thus, has been called the Anthropocene because humans are a "global forcing agent" (Zalasiewicz et al. 2010: 44). The effects include extinctions, invasive species, and climate change. This constitutes a modification of earlier classifications that posited the forcing agents to be geological, natural, or astronomical. At the same time, the term perhaps obscures the fact that humans have had a variety of quantitatively distinct impacts, and because of this, we need a finer-tuned model of human-mediated disturbance of natural environments (Isendahl 2010). Historical ecology has such a nuanced approach. Yet some researchers continue to regard ancient indigenous impacts on Amazonian biotic distributions, frequencies, and mass as negligible, or part of natural, expected processes of intermediate disturbance. They criticize use of univariate metrics such as species richness (i.e., diversity) rather than focusing on rare species. As such, they run the risk of perpetuating the myth that ancient cultural forests consist of common species, and only high forests are characterized by, or harbor, rare "forest-dependent" species (Barlow et al. 2011).

The case of the forests of the Beni, Bolivia (Llanos de Mojos) illustrates how humans impacted diversity upward but not through intermediate disturbance, but rather significant primary landscape transformation. Clark Erickson and I determined a cultural factor at work in order to explain the origins of tree species diversity at the mound site of Ibibate (Erickson and Balée 2006), and here, briefly, I recap that evidence. Ibibate is a *terra firme* forest of about 7 ha in extent; it is located on an anthropogenic mound measuring approximately 18 m in height at its highest point (Erickson 1995). At the base of the mound is a man-made ditch that maintains water year round; at least one causeway emanates from the mound and seems to cross the adjoining pampa at a distance of 4–5 km to the current Sirionó village of Ibibate, also located on a mound, though smaller in height and extent than Ibibate (Erickson 1995, 2006; Erickson and Balée 2006; Sington 2002). Two one-hectare inventories of forest were carried out in the environs of Ibibate mound, which is found within the Sirionó Indigenous Territory, about 40 km due east of the city of Trinidad. The first hectare of forest is located directly over the crest of the mound. It is 20×500 m in dimension with forty subplots of 10×25 m each; all trees $\geq$10 cm dbh (diameter at breast height) were collected (in 1993–1994) and identified, if not always to family, genus, and species, at least to morpho-species (see Campbell et al. 2006 on the validity of morpho-species as an analytic construct for describing diversity based on inventory data). There were 448 individual trees and woody vines in 55 species on

Table 2.1 Ten most dominant species, Ibibate mound

Species	Relative dominance
Attalea phalerata Mart ex. Spreng. (Arecaceae)	18.3
Gallesia integrifolia (Spreng.) Harms	15.35
Ampelocera ruizii Klotzsch (Ulmaceae)	10.96
Astrocaryum murumuru Mart. var. *murumuru*	9.27
Hura crepitans L. (Euphorbiaceae)	7.71
Ficus pertusa L.f. (Moraceae)	3.97
Calophyllum brasiliense Cambess. (Clusiaceae)	3.39
Sorocea guilleminiana Gaudich.	3.31
Dendropanax cuneatus (DC) Decne. & Planchon (Araliaceae)	3.07
Hirtella triandra Sw. subsp. *triandra*	2.9
Total	78.23

this plot. This measure of alpha diversity is high for the region, given the environmental gradients of latitude and rainfall (ca. 14° S latitude which is southerly and 1,520 mm/year, which is low for tropical moist forest). The ten most dominant species from Ibibate account for 78 % of the relative dominance of all 55 species on the plot; it is an "oligarchic" forest (Peters et al. 1989) insofar as it is heavily based on a few species, when seen in this perspective (Table 2.1).

Several of these species may also be found in seasonally flooded environments (such as *Hura crepitans*, *Astrocaryum murumuru*, and *Attalea phalerata*), and of these, *H. crepitans* is millions of years old; arguable vicariance accounts for the evolution in inundated environments of such species (Bush 1994), but they can also be located on the mound, where they are completely protected from river flooding, river avulsion, and other tectonic events as well as lateral channel migration otherwise common in areas of meandering rivers, which is what characterizes the lowland habitats of this area (Pärssinen and Korpisaari 2003). That is, evolved by vicariance in natural selection, and also selected for by historical, human activity. By way of comparison, and in a roughly equal time frame of landscape transformation, it has been recently found that from an inventory of old growth forest on top of the geoglyph known as Três Vertentes (Balée et al. in press), there were 149 species greater than or equal to 10 cm DBH per hectare. Geoglyphs are massive geometric formations of circles, squares, and rectangles; many of them have been uncovered for aerial viewing by deforestation (Pärssinen et al. 2009). The geoglyph of Três Vertentes, which is a gigantic circle larger than 1 ha in size, is also an oligarchic forest in the sense that the first ten most dominant species account for 71 % of the total dominance, and of these 10 species, two are palms (*Iriartea deltoidea* Ruiz & Pav., which is the most dominant species, and *Euterpe precatoria* Mart. var. *precatoria*) (Table 2.2). Although the geoglyph forest is much richer in species than Ibibate (at a total of 149 species vs. 55 per hectare), there are similarities between the two sites in terms of the cultural and historical *kinds* of species present. These are kinds of species that are affected by human movements and landscape transformations; their distributions are comprehensible in terms of historical ecology.

Table 2.2 Ten most dominant species, geoglyph Três Vertentes

Species	Relative dominance
Iriartea deltoidea Ruiz & Pav. (Arecaceae)	18.7
Tetragastris altissima (Aubl.) Swart.	12.2
Cecropia sp. 1 (Moraceae)	8.58
Acacia sp. 1 (Fabaceae)	7.29
Euterpe precatoria Mart. var. *precatoria* (Arecaceae)	5.82
Castilla ulei Warb. (Moraceae)	5.40
Brosimum lactescens (S.S. Moore) C.C. Berg (Moraceae)	5.01
Acacia polyphylla DC (Fabaceae)	3.05
Virola duckei A.C. Smith (Myristicaceae)	2.56
Celtis schipii Standl.	2.51
Total	71.12

The ancient people of Ibibate and environs evidently preferred to occupy the upland/wetland interface (Lombardo and Prümers 2010), and in this environment, they transformed the landscape. A second inventory at the base of the Ibibate mound, on the other side of one of the causeways leading out from the mound, was carried out in 1997. The dimensions of this plot were 100 × 100 m, and the area is seasonally flooded, unlike the site at the crest of the mound, which is never flooded. I had anticipated that species diversity here would be much lower (in terms of tree species, the lowland, flooded savanna that surrounds Ibibate is clearly lower in species diversity), as a transitional zone to the pampa (savanna). But at 425 individuals in 53 species, the diversity index is not statistically of significant difference from Ibibate. The two plots share 21 species, so the Jaccard coefficient (a/a + b) × 100, is 21/108 × **100** = 19.4. That is actually relatively high correspondence for adjacent tropical moist forest of different types (cf. Balée 1994: 134). But what is most important to grasp is that the current alpha diversity at the height of the mound would not exist had it not been for human intervention, the building of the mound in the first place, and that took place during a roughly 1,000 period ending roughly five or six hundred years ago (Erickson 1995); most of the mounds of the region of Ibibate appear to have been occupied during the period AD 400 and AD 1400 (as reviewed by Lombardo and Prümers 2010). It is quite possible that forest from the mound has spilled over onto the adjoining lowland area, expanding itself autochthonously at the expense of preexisting savanna (this hypothesis in regard to forests of the Baurés area to the north is argued in Erickson 2000); another possibility is that the seasonally flooded forest along the savanna margin, which like the forest on the mound itself, is rich in useful species (fruit trees, fuel species, and so on) that could have been used and protected by the ancient inhabitants of the mound. Barlow et al. (2011) argue that conservation should not be focused on common species in secondary forests as the result of human action, but rather on "forest-dependent" species. What they seem to miss is that perhaps some of those species are dependent in historical fact on human activity that in the past built forests. Actually, protection of some species on the mound has occurred in the recent past, by the Sirionó themselves. Indeed, some of the species on the height of the mound do not tolerate flooding and are not encountered, according to knowledgeable Sirionó elders, ever outside the boundaries of the high mounds. One of these is *turumbúri* tree (*Sorocea guilleminiana* Gaudich., mulberry family), used in making a ceremonial, fermented beverage of the Sirionó used in important rituals (Balée 2000). Arguably, the tree *turumbúri* is a rare and endemic species: Namely, it is endemic to areas disturbed by humans.

2.3 Conclusion

The various models of diversity can be interwoven to some extent with the research program of historical ecology. The timelines in the case of Ibibate can be entertained in terms of evolution of phyla, adaptation to environmental gradients, and historical factors of human disturbance. As to species evolution, some species extant on Ibibate existed millions of years ago, probably in the context of the original formation of the Amazon River Basin itself, following the Andean orogeny. Species diversity in Amazonian forests is also obviously limited by synchronic, environmental gradients of latitude and rainfall. Other timelines are much more recent. It is possible that some of the species on the Ibibate mound migrated out of centers of endemism from elsewhere in the Amazon by principles related to the refuge model, though there is little endemism in the area as a whole, either in undisturbed or transformed landscapes. Floristic diversity (which usually is associated with faunal diversity, as the refuge theory holds) is nevertheless high for the area (a wetland savanna) in general. The calculation of the origins of this diversity takes into account speciation events at millions of years ago, possible speciation events at tens of thousands of years ago (à la a modified form of the refuge theory), environmental gradients (in this case, latitude and rainfall), and human history, occupation, and development of the area within the past two thousand years. This history is not a case of intermediate disturbance, which Barlow et al. (2011) proposed as the only method by which humans can increase alpha diversity. Rather, it is a case of *primary landscape transformation*: a complete upheaval of species and replacement of these by different species. That is, grasses and sedges were replaced by trees and lianas on Ibibate, due to human influence. The human history of the area involved a built environment (the mound) that permitted the growth and maintenance of a *terra firme* tropical moist forest, and the alpha diversity of that forest can only be understood once the human factor is taken into account. This is probably the case with many forests of Amazonia as yet unstudied. There are still geoglyphs, for example, covered in forest (Pärssinen et al. 2009), though perhaps not for long, given the velocity of habitat fragmentation in eastern Acre and environs.

Historical ecology therefore admits of varying timelines in the total explication of diversity, but the one indispensable feature is human activity. That is because human activity accounts for the distribution patterns observed in the present at the alpha level and perhaps also at the beta level (when considering the addition of the forest on the mound/savanna margin). For Amazonia, then, the timeline of diversity focuses on three general reference frames: millions of years ago (Miocene), tens of thousands of years ago (Pleistocene), and hundreds to thousands of years ago (Holocene, including the human, historical presence). This approach to time and diversity in Amazonia, which breaks time down into significant segments relating to the origins of species diversity, may be understood to be a working model within the framework of historical ecology.

Appendix

Viveiros de Castro (1996) uses the term "standard" model of Amazon ethnology to refer to what I am calling the adaptationist model.

Recent evidence posits that Incan civilization, in a military if not also economic and cultural sense, did indeed penetrate and influence Amazonian prehistory, at least in the upper Amazon (Pärssinen and Korpisaari 2003).

I mean "research program" in the sense of Lakatos (1980) and would distinguish it from the "paradigm" concept of Kuhn (1970) (though I did not do so originally—Balée 1998). The reason for the distinction is historical ecology is probably not a paradigm (cf. Biersack 1999: 8–9), since paradigms demand overwhelming consensus in the scientific community, and all essential problems (in this case, research problems concerning humans and the environment) need to have their own models of explication and deduction that originate in the axioms of the paradigm. Such consensus does not yet exist in historical ecology. The term research program is less rigid and more appropriate to the notion of historical ecology, allowing as it does for less consensus but a relatively widely connected body of research, and does exist in historical ecology (e.g., Crumley 1994, 2001; Balée 1998, 2006).

References

Balée W (1989) The culture of Amazonian forests. In: Posey DA, Balée W (eds) Resource management in Amazonia: indigenous and folk strategies, Adv Econ Bot 7. New York Botanical Garden, Bronx, pp 1–21

Balée W (1994) Footprints of the forest: Ka'apor ethnobotany—the historical ecology of plant utilization by an Amazonian people. Columbia University Press, New York

Balée W (1998) Historical ecology: premises and postulates. In: Balée W (ed) Advances in historical ecology. Columbia University Press, New York, pp 13–29

Balée W (2000) Elevating the Amazonian landscape. For Appl Res Publ Policy 15(3):28–33

Balée W (2006) The research program of historical ecology. Ann Rev Anthropol 35:75–98

Balée W (2010) Contingent diversity on anthropic landscapes. Diversity 2:163–181

Balée W, Erickson CL (eds) (2006) Time and complexity in historical ecology: studies in the neotropical lowlands. Columbia University Press, New York

Balée W, Schaan DP, Whitaker, JA, Holanda R (2014) Florestas antrópicas no Acre: Inve ntário florestal do geoglifo Três Vertentes, Acrelândia, AC. Amazônica (in press)
Barlow J, Gardner TA, Lees AC, Parry L, Peres CA (2011) How pristine are tropical forests? An ecological perspective on the pre-Columbian human footprint in Amazonia and implications for contemporary conservation. Biol Conserv 151:45–49. doi:10.1016/j.biocon.2011.10.013
Bates J (2001) Avian diversification in Amazonia: evidence for historical complexity and a vicariance model for a basic diversification pattern. In: Vieira ICG, Cardoso da Silva JM, Oren DC, D'Incao MA (eds) Diversidade biológica e cultural da Amazônia. Museu Paraense Emílio Goeldi, Belém, pp 119–137
Begon M, Harper JL, Townsend CR (1990) Ecology: individuals, populations and communities, 2nd edn. Blackwell, Boston
Biersack A (1999) Introduction: from the "new ecology" the new ecologies. Amer Anthropol 101:5–18
Braudel F (1993) A history of civilizations. Trans. R. Mayne. Penguin, New York
Brondizio ES, Siqueira AD (1997) From extractivists to forest farmers: changing concepts of agricultural intensification and peasantry in the Amazon estuary. Res Econ An 18:233–279
Brondizio ES, McCracken S, Moran E, Siqueira A, Nelson D, Rodriguez-Pedraza C (2002) The colonist footprint: toward a conceptual framework of land use and deforestation trajectories among small farmers in the Amazonian frontier. In: Wood CH, Porro R (eds) Deforestation and land use in the Amazon. University of Florida Press, Gainesville, pp 133–161
Bush MB (1994) Amazonian speciation: a necessarily complex model. J Biogeogr 21:5–17
Bush MB, Silman MR, de Toledo MB, Listpad C, Gosling WD, Williams C, de Oliveira PE, Krisel C (2007) Holocene fire and occupation in Amazonia: records from two lake districts. Phil Trans R Soc B 362:209–218
Campbell DG, Ford A, Lowell KS, Walker J, Lake JK, Ocampo-Raeder C, Townesmith A, Balick M (2006) The feral forests of the Eastern Petén. In: Balée W, Erickson CL (eds) Time and complexity in historical ecology: studies in the neotropical lowlands. Columbia University Press, New York, pp 21–55
Clement CR (1999a) 1492 and the loss of Amazonian crop genetic resources I. The relation between domestication and human population decline. Econ Bot 53:188–202
Clement CR (1999b) 1492 and the loss of Amazonian crop genetic resources II: crop biogeography at contact. Econ Bot 53:203–216
Clement CR, Junqueira AB (2010) Between a pristine myth and an impoverished future. Biotropica 42:534–536
Colinvaux P (1987) Amazon diversity in the light of the paleoecological record. Quat Sci Rev 6:93–114
Colinvaux P, Oliveira PE, Bush MB (2000) Amazonian and neotropical plant communities on glacial time-scales: the failure of the aridity and refuge hypotheses. Quat Sci Rev 19:141–169
Crumley CL (ed) (1994) Historical ecology: cultural knowledge and changing landscapes. School of American Research Press, Santa Fe
Crumley CL (ed) (2001) New directions in anthropology and environment: intersections. Altamira Press, Walnut Creek
Daly D, Silveira M (2002) Aspectos florísticos da bacia do Alto Juruá: História botânica, peculiaridades, afinidades e importância para a conservação. In: da Carneira Cunha M, de Barbosa Almeida M (eds) Enciclopédia da floresta. Companhia das Letras, São Paulo, pp 53–63
Denevan WM (1992) The pristine myth. Ann Assoc Am Geogr 82:369–385
Denevan WM (2001) Cultivated landscapes of native Amazonia and the Andes. Oxford University Press, New York
Earle TK (2002) Bronze Age economics: the beginnings of political economies. Westview Press, Boulder
Erickson CL (1995) Archaeological perspectives on ancient landscapes of the Llanos de Mojos in the Bolivian Amazon. In: Stahl P (ed) Archaeology in the American tropics: current analytical methods and applications. Cambridge University Press, Cambridge, pp 66–95

Erickson CL (2000) An artificial landscape-scale fishery in the Bolivian Amazon. Nature 408:190–193
Erickson CL (2006) The domesticated landscapes of the Bolivian Amazon. In: Balée W, Erickson CL (eds) Time and complexity in historical ecology: studies in the Neotropical lowlands. Columbia University Press, New York, pp 235–278
Erickson CL (2008) Amazonia: the historical ecology of a domesticated landscape. In: Silverman H, Isbell WH (eds) Handbook of South American archaeology. Springer, New York, pp 157–183
Erickson C, Balée W (2006) The historical ecology of a complex landscape in Bolivia. In: Balée W, Erickson C (eds) Time and complexity in historical ecology: studies in the neotropical lowlands. Columbia University Press, New York, pp 187–233
Fairhead J, Leach M (1996) Misreading the African landscape: society and ecology in the forest-savanna mosaic. Cambridge University Press, Cambridge
Fisher W (2000) Rain forest exchanges: industry and community on an Amazonian frontier. Smithsonian Institution Press, Washington DC
Gentry A (1988) Changes in plant community diversity and floristic composition on environmental and geographical gradients. Ann Mo Bot Gard 75:1–34
Haffer J (1969) Speciation in Amazonian forest birds. Science 165:131–137
Haffer J (2001) Hypotheses to explain the origin of species in Amazonia. In: Oren DC, D'Incao MA, Vieira I, da Cardoso Silva JM (eds) Diversidade biológica e cultural da Amazônia. Museu Paraense Emílio Goeldi, Belém, pp 45–118
Heckenberger M (2006) History, ecology, and alterity: visualizing polity in ancient Amazonia. In: Balée W, Erickson CL (eds) Time and complexity in historical ecology: studies in the neotropical lowlands. Columbia University Press, New York, pp 311–340
Heckenberger M, Russell JC, Fausto C, Toney JR, Schmidt MJ, Pereira E, Franchetto B, Kuikuro A (2008) Pre-Columbian urbanism, anthropogenic landscapes, and the future of the Amazon. Science 321:1214–1217
Hornborg A, Crumley C (eds) (2007) The world system and the earth system: global socioenvironmental change and sustainability since the Neolithic. Left Coast Press, Walnut Creek
Isendahl C (2010) The Anthropocene forces us to reconsider adaptationist models of human-environment interactions. Environ Sci Technol 44:6007
Kuhn T (1970) The structure of scientific revolutions, 2nd edn. University of Chicago Press, Chicago
Lakatos I (1980) The methodology of scientific research programmes (Philosophical Papers, volume 1). Cambridge University Press, Cambridge
Lombardo U, Prümers H (2010) Pre-Columbian human occupations in the eastern plains of the Llanos de Moxos, Bolivian Amazonia. J Archaeol Sci 3:1875–1885
Lovejoy N, Bermingham RN, Martin AP (1998) Marine incursions into South America. Nature 396:421–422
Mann C (2002) 1491. Atlantic Mon 289(3):41–53
Mann C (2005) 1491: new revelations of the Americas before Columbus. Afred A Knopf, New York
Marsh GP (1885) [orig. 1864] The earth as modified by human action. In: Of man and nature, 2nd edn (revised). Scribner, New York
McEwan C, Barreto C, Neves EG (eds) (2001) Unknown Amazon: culture in nature in ancient Brazil. The British Museum Press, London
McGrane B (1989) Beyond anthropology: society and the other. Columbia University Press, New York
McMichael CH, Piperno DR, Bush MB, Silman MR, Zimmerman AR, Raczka MF, Lobato LC (2012) Sparse pre-Columbian human habitation in western Amazonia. Science 336:1429–1431
Meggers BJ (1996) Amazonia: man and culture in a counterfeit paradise, Revised edn. Smithsonian Institution, Washington DC
Moran EF (2000) Human adaptability: an introduction to ecological anthropology, 2nd edn. Westview, Boulder

Mörner N-A, Rossetti DF, Toledo PM (2001) The Amazon rainforest: only some 6–5 million years old. In: Oren DC, D'Incao MA, Vieira ICS, da Cardoso Silva JM (eds) Diversidade biológica e cultural da Amazônia. Museu Paraense Emílio Goeldi, Belém, pp 3–18

Nelson BW, Ferreira CAC, Da Silva MF, Kawasaki ML (1990) Endemism centres, refugia and botanical collection density in Brazilian Amazonia. Nature 345:714–716

Pärssinen M, Korpisaari A (eds) (2003) Western Amazonia—Amazônia Ocidental: multidisciplinary studies on ancient expansionistic movements, fortifications and sedentary life. Revall Institute for Area and Cultural Studies, University of Helsinki, Helsinki

Pärssinen M, Schaan DP, Ranzi A (2009) Pre-Columbian geometric earthworks in the upper Purus: a complex society in western Amazonia. Antiquity 83(322):1084–1095

Peters CM, Balick MJ, Kahn F, Anderson AB (1989) Oligarchic forests of economic plants in Amazonia: utilization and conservation of an important tropical resource. Conserv Biol 3:341–349

Raffles H (2002) Amazonia. Princeton University Press, Princeton

Räsäsen ME, Linna AM, Santos JCR, Negri FR (1995) Late Miocene tidal deposits in the Amazonian foreland basin. Science 269:386–390

Rosenzweig ML (1995) Species diversity in space and time. Cambridge University Press, New York

Sington D (Director) (2002) BBC horizon film: the secret of El Dorado

Stahl P (2002) Paradigms in paradise: revising standard Amazonian prehistory. Rev Archaeol 23(2):39–51

Viveiros de Castro E (1996) Images of nature and society in Amazonian ethnology. Ann Rev Anthropol 25:179–200

Wallerstein I (1974) The modern world-system: capitalist agriculture and the origins of the European world-economy in the sixteenth century. Academic Press, New York

Whitaker RA (1972) Evolution and measurements of species diversity. Taxon 2:213–251

Whitehead N (1998) Ecological history and historical ecology: diachronic modeling versus historical explanation. In: Balée W (ed) Advances in historical ecology. Columbia University Press, New York, pp 30–41

Whitehead N (2003) Introduction. In: Whitehead NL (ed) Histories and historicities in Amazonia. University of Nebraska Press, Lincoln, pp vii–xx

Whitmore TC (1990) An introduction to tropical rain forests. Clarendon Press, London

Wolf E (1982) Europe and the people without history. University of California Press, Berkeley and Los Angeles

Zalasiewicz J, Williams M, Steffen W, Crutzen P (2010) The new world of the Anthropocene. Environ Sci Technol 44:2228–2231

Chapter 3
Phylogenetic Diversity and the Sustainable Use of Biodiversity

Daniel P. Faith and Laura J. Pollock

Abstract Sustainable use of biodiversity requires the use of biodiversity in a way that does not foreclose benefits for future generations. Biodiversity option values reflect this capacity to provide future benefits that are often unanticipated. The phylogenetic diversity measure, PD, quantifies the option values represented by different sets of species. PD can be interpreted as counting-up features of species. This allows species-level ecological indices to be converted to phylogenetic indices, including PD complementarity and PD endemism, and integrated into systematic conservation planning. PD's power law relationship with species counts supports findings that initial species losses may retain high PD. This suggests that occasional loss of current-use species might not reduce overall PD. However, if species that are currently useful to society are concentrated in particular clades on the phylogeny, then their loss may imply high-PD loss. Conservation of current-use species can maintain overall PD and option values. However, systematic conservation planning results suggest that conservation of phylogenetically clumped current-use species, within a given conservation budget, can produce a tipping point in which the capacity to retain high-PD collapses.

3.1 Introduction

In this chapter, we will link one of the most fundamental aspects of biodiversity—the tree of life or phylogeny—to one of the most practical concerns of biodiversity conservation—the sustainable use of biodiversity. This topic contributes another

---

D. P. Faith (✉)
The Australian Museum Sydney, Sydney 2010, Australia
e-mail: danfaith8@yahoo.com.au

L. J. Pollock
National Environmental Research Program, School of Botany,
The University of Melbourne, Victoria 3010, Australia
e-mail: laurajs@unimelb.edu.au

L. M. Verdade et al. (eds.), *Applied Ecology and Human Dimensions in Biological Conservation*, DOI: 10.1007/978-3-642-54751-5_3,

perspective to our book's overall theme on new directions for integrating applied ecology, human dimensions, and biological conservation. A precursor for this book was the 2009 Biota-FAPESP international workshop on "Applied ecology and human dimensions in biological conservation" (http://www.fapesp.br/5434). The workshop highlighted various new strategies in applied ecology, associated with emerging stronger links to human dimensions and to historical perspectives. We will touch on these themes in exploring how phylogeny helps us to understand and achieve sustainable use of biodiversity.

It is timely to consider the challenges of sustainable use of biodiversity. During 2012, the United Nations Conference on Sustainable Development (UNCSD or "Rio+20") was held in Brazil, marking 20 years since the original conference that gave birth to the convention on biological diversity (CBD). The major outcome document from the UNCSD conference refers frequently to "sustainable use of biodiversity" (UNCSD 2012). However, nearly all the references are part of a general call for "the conservation and sustainable use of biodiversity". This invites some fresh consideration about how conservation and sustainability goals are linked.

Article 2 of the CBD (http://www.cbd.int/convention/articles/?a=cbd-02) defines "sustainable use of biodiversity" as:

> the use of biological diversity in a way and at a rate that does not lead to the long-term decline of biological diversity, thereby maintaining its potential to meet the needs and aspirations of present and future generations.

Thus, sustainable use of biodiversity presents the challenge of providing uses to satisfy current needs while maintaining the capacity to anticipate and satisfy the needs of future generations, through other uses.

The idea that sustainable use of biodiversity requires consideration of possible future needs of society echoes the earliest justifications for biodiversity conservation, based on the idea of future prospective human uses. For example, the World Conservation Strategy (IUCN 1980) called for conservation of diversity "for present and future use". McNeely (1988) referred to biodiversity conservation as providing a "safety net of diversity" based on its "option values" (see also Reid and Miller 1989). Option values of biodiversity are the biodiversity values that provide benefits and uses, often unanticipated, for future generations. This link means that measures of biodiversity—under the standard definition of living variation across genes, species, and ecosystems—can be interpreted as measures of option values (for review and discussion, see Faith 2012a, b, 2013).

Preservation of these biodiversity option values arguably is central to any real sustainable use programme. However, option values sometimes are underappreciated in current debates about biodiversity conservation (Faith 2013). When human benefits are discussed, the term "ecosystem services" typically is used as a catch-all to cover any benefits from ecosystems that range from pristine to heavily human-modified. While ecosystem services consequently might include anything and everything, actual ecosystem services case studies typically have emphasized well-known human uses and benefits, rather than possible future uses that are currently unknown (for discussion, see Faith 2010).

The focus on current essential ecosystem services also is apparent in the new Strategic Plan and 2020 Aichi targets of the CBD (www.cbd.int/doc/strategic-plan/2011-2020/Aichi-Targets-EN.pdf). These new targets provide a mixed message about the importance of benefits from biodiversity for future generations. The mission of the Strategic Plan is "to take effective and urgent action to halt the loss of biodiversity in order to ensure that by 2020 ecosystems are resilient and continue to provide essential services". A related Aichi target calls for maintenance of ecosystems that provide "essential" services. This phrasing may encourage conservation actions that focus on continued supply of those services known to be essential now, rather than worrying about future services and uses that are presently unknown and unanticipated. This same issue extends to other Aichi targets. For example, another target refers to preservation of genetic diversity, but the stated focus is on known crop species and their close relatives.

Recent characterizations of "biodiversity" reflect this popular focus on current uses. "Biodiversity" is interpreted primarily as a foundation for current uses, and biodiversity conservation is sometimes seen as accomplished by ecosystem services conservation. For example, Perrings et al. (2010) suggested that "what and how much biodiversity should be targeted for conservation depends on what services are important" (for discussion, see Faith 2011). When "important" services define the biodiversity of interest in this way, the adopted definitions and measures of biodiversity may simply re-express services in terms of their ecological basis (such as abundance and species' interactions). Traditional definitions of biodiversity recently have been expanded to include many of these aspects of species-level ecology (for discussion, see Faith 2011, 2013). For example, one ecosystem services study (Díaz et al. 2009) considered biodiversity as "the number, abundance, composition, spatial distribution, and interactions of genotypes, populations, species, functional types and traits, and landscape units in a given system". These ecological aspects may be important to the analysis of current ecosystem services, but may not help quantify option values.

Consideration of biodiversity option values sometimes is seen as less practical than strategies that link the biodiversity of interest to the ecology of ecosystem services. For example, Mace et al. (2010) argued that "to maintain biodiversity so as not to foreclose the options open to future generations… would entail a goal of no overall loss of biodiversity… we suggest this is unlikely to be achievable". Others have neglected biodiversity option values, even when they do acknowledge biodiversity as something distinct from ecosystem services. In such cases, biodiversity may be characterized as primarily all about intrinsic (non-anthropocentric) values, with the human uses largely captured by the ecosystem services (for discussion, see Faith 2012a, b). In contrast, the UNCSD outcome document (UNCSD 2012) did state the importance of biodiversity values extending beyond intrinsic values:

> We reaffirm the intrinsic value of biological diversity, as well as the ecological, genetic, social, economic, scientific, educational, cultural, recreational and aesthetic values of biological diversity and its critical role in maintaining ecosystems that provide essential services…

However, one limitation of this affirmation is that it did not explicitly highlight potential future uses or option values of biodiversity.

Some of these differences in perspective may be matter of definition (see Redford and Richter 1999). The Millennium Ecosystem Assessment (MA) (2005) has provided helpful guidelines, by distinguishing between ecosystem services and biodiversity and by highlighting the option values of biodiversity. The MA noted that "a general lesson is that poor measurement of biodiversity reduces the capacity to discover and implement good trade-offs and synergies between biodiversity and ecosystem services". The MA also concluded that "sometimes responses to this information problem may… neglect the difficult problem of finding surrogates for global option values".

Progress on this surrogates problem would help establish clear links between option values and sustainable use of biodiversity and would complement both the current-use and the intrinsic-value perspectives on biodiversity. We suggest that solutions to the "difficult" problem of finding surrogates for biodiversity option values depend on effective quantification of living variation. Such measures should include the variation among species in characters or features, because these are elements of biodiversity that might correspond to future uses and benefits (Faith 1992a, 2013). Here, we consider phylogeny, or the tree of life, as a basis for making inferences about biodiversity at this level of features of species. Our premise is that greater phylogenetic diversity, or feature diversity, implies greater option values—a greater number of potential future uses and benefits. Thus, phylogeny has particular relevance to sustainable use of biodiversity.

We consider a specific phylogenetic diversity measure PD (Faith 1992a, b) as our measure of feature diversity and option values. The PD measure not only allows us to talk about future uses but also can integrate information about current uses. Conservation of species that are currently used provides some level of conservation of the phylogenetic diversity and option values of the corresponding taxonomic group (e.g. legumes; see below). Therefore, the conservation of currently used species partially satisfies the requirements for sustainable use of biodiversity. However, a theme of this chapter is that there are advantages in integrating or balancing the conservation investments in known current-use species with conservation of broader phylogenetic diversity. We suggest that overemphasizing species that are currently valuable could reduce our capacity to preserve these broader option values—potential future uses—represented by the phylogenetic diversity within a given taxonomic group.

Our chapter is structured as follows. First, we review the phylogenetic diversity measure (PD) and the links from phylogenetic diversity to option values. Here, we highlight the need to look at gains and losses of features and PD, not just overall PD values. We show how we can replace many standard indices of species-level ecology with phylogenetic indices that count up features, not species. Second, we describe how PD calculations are relevant to the problem of conservation and sustainable use of biodiversity. Here, we describe the fundamental relationship between PD and number of species, and how this relationship changes depending on the pattern of species gains and losses across the phylogeny (i.e. phylogenetically dispersed or clumped). Third, we describe conservation planning and decision-making that integrates PD, current uses, and additional factors such as costs of conservation. We examine conservation planning scenarios using

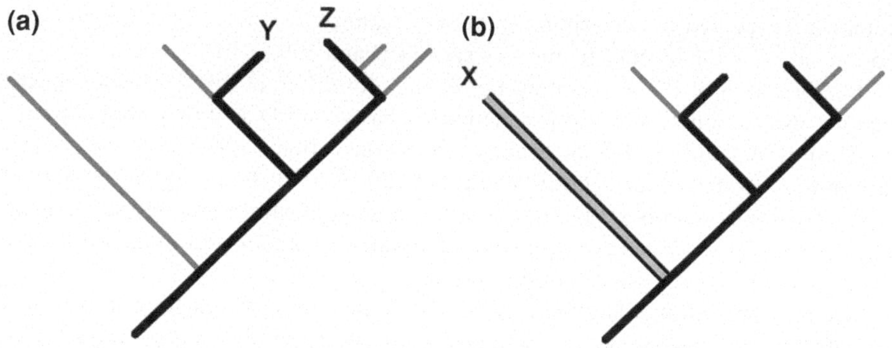

Fig. 3.1 Hypothetical phylogenetic trees illustrating PD. **a** The PD represented by the set of two species, *Y* and *Z*, as *darker lines*. **b** Addition of species *X* increases the PD by the amount shown by the *double-line* segment. This additional length needed to arrive at *X* is the PD complementarity value of *X*

this framework. We explore the contribution of conservation of current-use species to conservation of PD and conclude that conservation of currently used species should be complemented by direct PD conservation. We finish by returning to the general call for "the conservation and sustainable use of biodiversity" by describing how these two goals should be interlinked through conservation planning.

3.2 PD, Feature Diversity, and a Calculus of Option Values

3.2.1 Evosystem Services and PD

One limitation of the ecosystem services framework is that it is very place-based in focusing on processes within ecosystems as the basis for human benefits. A complementary perspective can focus more on evolutionary processes (Faith et al. 2010; Hendry et al. 2010). Evolutionary processes, as reflected in the tree of life, generate benefits provided by characteristics or features of species. These current and future benefits for humans have been referred to as evolutionary or "evosystem services" (see Faith et al. 2010).

The phylogenetic diversity measure, "PD" (Faith 1992a, b), helps us to quantify these current and potential future benefits derived from the tree of life. The PD of a given set of species is defined as the minimum total length of all the phylogenetic branches required to connect all those species on the tree (Fig 3.1a). PD provides a natural way to talk about future uses and benefits provided by species because the counting-up of branch lengths links sets of species to their expected relative diversity of characters or features. PD is based on a standard model of evolutionary process that implies that shared ancestry should account for shared features (Faith 1992b). Therefore, any subset of species that has greater phylogenetic diversity, PD, will represent greater feature diversity. Because larger PD values are expected to correspond to greater feature diversity, PD values indicate option values at the level of features of species (Faith 1992a, b).

Interpretation of PD as counting-up features for different sets of species means that we also can interpret various calculations based on PD as if they are counting-up features. A family of PD measures extends conventional species-level measures and indices to the features level (Faith and Baker 2006; Faith 2008a; Nipperess et al. 2010). For example, PD dissimilarities among localities are calculated using phylogenetic tree branches, producing measures analogous to standard Bray–Curtis and other species-level dissimilarities.

Because PD implicitly counts features among sets of species, it provides straight-forward measures of complementarity (i.e. number of additional features gained or lost) and endemism (i.e. number of features unique to a species or to an area). Complementarity and endemism values can be calculated for species or for areas. Priority setting for conservation then may focus, for example, on the PD loss if a threatened species is pruned by extinction from the phylogenetic tree (Faith 1992a, 1994). The magnitude of the PD loss from loss of any one species naturally depends on the fate of its close relatives. The loss could be large if the species were the only remaining survivor in a highly distinctive group (on the basis of PD complementarity; Fig 3.1b). Examples of PD complementarity calculations are found in Forest et al. (2007) and Faith and Baker (2006). Faith (1994) provides examples of PD endemism, including PD endemism of amphipods for northwest Tasmania (see also Faith et al. 2004).

The PD measure is now recognized as a basis for setting conservation priorities among species or areas (Faith 1992a; Forest et al. 2007; Mace 2003). Bordewich and Semple (2012) state that "phylogenetic diversity (PD) has emerged as a leading measure in quantifying the biodiversity of a collection of species". Davies and Buckley (2011) conclude that "The loss of PD, quantified in millions of years, provides a resonant symbol of the current biodiversity crisis".

While priorities properly focus on PD gains and losses, it is sometimes assumed that the total PD of a locality is the basis for priority setting (e.g. Isambert et al. 2011). In fact, PD complementarity and endemism are critical to such planning (Faith 1992a; Faith et al. 2004). For example, PD complementarity is now recognized as useful for conservation planning based on molecular trees from DNA barcoding (Faith and Baker 2006). Krishnamurthy and Francis (2012) review the use of PD and DNA barcoding in conservation. In this context, Smith and Fisher (2009) document how PD calculations are important in providing robust estimates of complementarity values. We return to PD and conservation planning below.

3.2.2 *Phylogenetic Patterns of Current Uses*

Some applications of PD have explicitly referred to feature diversity and option values. Examples include applications to bioprospecting, where greater PD indicates greater potential for novel discoveries (Pacharawongsakda et al. 2009, see also Saslis-Lagoudakis et al. 2011). Similarly, a study of bioprospecting of piscine venoms (Smith and Wheeler 2006) stressed the utility of phylogeny in providing predictions about unknown characteristics of species (see also Tulp and Bohlin 2002). However, the actual success of phylogeny, and the PD measure, in capturing future uses has had little investigation.

Forest et al. (2007) have provided some evidence for the utility of PD as for quantifying estimated feature diversity and option values. They examined the distribution of angiosperm plants with known human uses (classified as medicinal, food, and all other uses) on an estimated phylogenetic tree for nearly 900 genera found in the Cape hotspot of South Africa. Their information source, the Survey of Economic Plants for Arid and Semi-Arid Lands (SEPASAL), reports on the uses of tropical and subtropical wild and semi-domesticated plants. Forest et al. (2007) labelled a given genus as "useful" if it had at least one species found in the Cape and recorded in this database.

Forest et al. (2007) first asked how each use-type was distributed phylogenetically. They found that that each use-type was clumped on the tree: common ancestry often could account for taxa with the same use. This pattern suggests that phylogeny may help predict useful species, at least within any one use-type. This result corresponds to other findings. For example, Saslis-Lagoudakis et al. (2011) have found similar phylogenetic clumping for some use categories in legumes. However, Forest et al. (2007) also found that preserving species of one use-type did not do a good job of protecting species of another use-type. They found that knowledge of which plants were useful in one category would not be a good predictor of which plants were useful under another category. This suggests that protecting species with known uses generally would not be an adequate way to protect species with yet-to-be discovered uses. Forest et al. (2007) also determined that PD was the best general predictor over different use-types. Their conclusion was that current uses would not predict taxa with future uses, but that conservation of PD may effectively preserve options for the future.

These phylogenetic predictions about current and future uses highlight the role of phylogeny in capturing option values. To the extent that we are focusing on one use-type, good guesses might be made about which other species provide that use-type, based on any phylogenetic clumping of that use. Predicting which species might generally be "useful" in any of a variety of ways is more difficult. We agree with the conclusions of Beattie et al. (2011) that "the benefits of bioprospecting have emerged from such a wide range of organisms and environments worldwide that it is not possible to predict what species or habitats will be critical to society, or industry, in the future". As illustrated in the Forest et al. (2007) study, over a wide range of uses, it is not possible to predict which species will be useful. On the other hand, it is possible to increase our chances that a future useful species will still exist—we can do that by maximizing conservation of phylogenetic diversity. Thus, we shift the goals of prediction away from specific instances to predictions about the relative amount of option value retained by different subsets of persisting species.

If conservation of PD is desirable as a way to preserve option values, then to some extent this could be achieved by retaining lots of species. In the next section, we discuss the fundamental curve linking species number to PD. We also explain why in decision-making we are interested in departures from this general curve and how this is facilitated by integrating PD into conservation planning.

3.3 Departures from a Basic PD: Species Relationship

3.3.1 Phylogenetically Clumped or Dispersed Species Losses

The PD of a set of species will generally increase as more species are added to the set, and it is sometimes argued that conservation priorities based on maximizing species richness will also ensure conservation of phylogenetic diversity (e.g. Rodrigues and Gaston 2002). It is important therefore to consider the relationship between species number and PD and how it varies. Faith (2008a) proposed a power law curve for the PD–species relationship (see also Faith and Williams 2006):

> The total PD represented by different-sized sets of taxa defines a "features/taxa" curve, analogous to the well-known species/area curve. Random taxon samples of different sizes from a given phylogenetic tree produced a roughly linear relationship in log–log space.

Morlon et al. (2011) provided empirical support for this proposed power law model, based on estimated PD–species curves for four phylogenetic trees from four Mediterranean-type ecosystems. For each value of species richness (S), they calculated the PD of 100 communities obtained by randomly sampling S species from the phylogeny. This process revealed a power law PD–species relationship for all four phylogenies. This relationship is linear in log–log space (Fig. 3.2).

This relationship reveals some possible implications of species gains and losses on conservation of PD. One is that initial losses of species may mean only small losses in PD. At the other end of the curve, initial gains in protected species can mean large gains in PD. As the size of the protected set grows larger, the rate of gain in PD becomes progressively lower.

Those are expected patterns for the basic PD–species relationship—found when the number of species varies through random selection of species from the phylogeny. Real-world losses (and gains) will be non-random. Several studies have examined patterns of loss of phylogenetic diversity for a given number of species extinctions (reviewed in Morlon et al. 2011). The amount of actual PD loss depends in part on whether species extinctions are clumped or well-dispersed on the phylogenetic tree. For example, several studies looking at climate change impacts suggest relatively small PD losses (e.g. Yesson and Culham 2006). The climate change impacts spread out over the phylogenetic trees mean that deeper branches throughout the tree have at least one surviving descendent. Thuiller et al. (2011) similarly found small PD loss given dispersed species losses on the phylogenies for three different taxonomic groups.

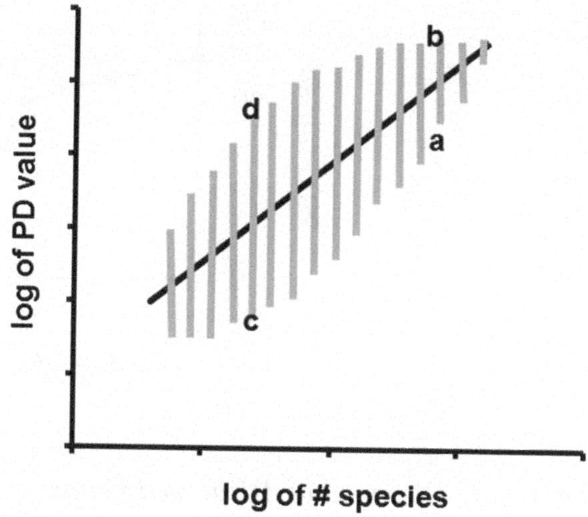

Fig. 3.2 A schematic diagram illustrating the power curve relationship between PD and number of species. In log–log space, this relationship is a straight line (*dark line in plot*). The power curve is produced by average PD values for random sets of species of a given size. Non-random sets will produce higher- or lower-PD values. The *grey bars* represent the range of possible values of PD for each number of species. *Points a* and *c* illustrate possible low-PD outcomes, and *points b* and *d* illustrate possible high-PD outcomes

In contrast, some studies have found that species extinction is concentrated on the phylogeny ("clumped"), resulting in a disproportionate loss of PD. One cause of such disproportionate loss is the occurrence of entire clades in the same threatened location. For example, this kind of clumping accounts for Huang et al.'s (2012) finding that several biodiversity hot spots in southern Asia and Amazonia are likely to lose "an unexpectedly large proportion of PD".

Faith et al. (2010) describe phylogenetically clumped impacts as a "tipping point" problem. Successive species extinctions each may imply only a moderate loss of PD, until the last descendent species from a long branch goes extinct, and the long branch representing a large amount of PD is now lost (Fig. 3.3). They advocated a form of "phylogenetic risk analysis" (Faith 2008b) to guide conservation decisions that try to reduce the risk of these worst case losses, or "tipping point" outcomes.

Thus, while random losses of species initially produce values near the top of the line in Fig. 3.2, non-random losses can produce markedly different results. Phylogenetically clumped losses may result in lower-PD outcomes (Fig. 3.2, point a), while phylogenetically well-dispersed losses could result in higher-PD outcomes (Fig. 3.2, point b). These scenarios may be relevant to the sustainable use of biodiversity. If current uses are phylogenetically clumped, as found in the study of Forest et al. (2007), then the loss of those species could imply a large PD loss.

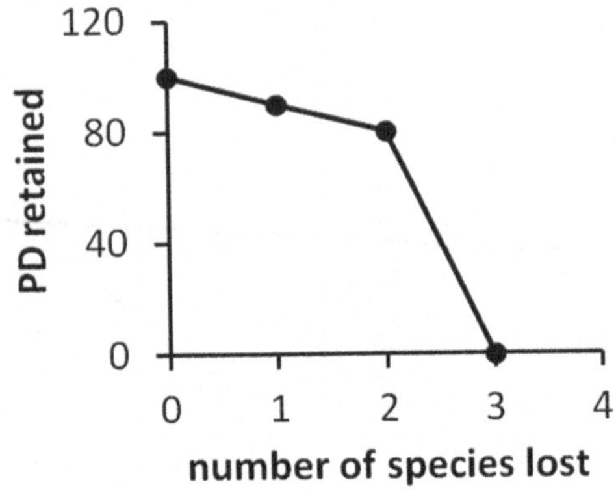

Fig. 3.3 The plot shows the PD retained as species are lost from a portion of a hypothetical phylogenetic tree having a long branch, leading to three closely related species. Loss of 1 or 2 species implies only small PD loss, but loss of the third species also means loss of the long ancestral branch

3.3.2 Phylogenetically Clumped or Dispersed Gains in Species Conservation

We considered a scenario above where current use of a species might lead to its loss (through some form of overuse). However, identified current uses of elements of biodiversity naturally also may act as an incentive for the conservation of those elements of biodiversity. For example, Penafiel et al. (2011) reviewed the literature on the contribution of plant and animal species to human diets and found that local food biodiversity is an important contributor of nutritious diets. They concluded that the use of this variety of species in the diet has promoted the conservation of this food biodiversity.

Conservation of a set of current-use species may or may not imply the preservation of lots of PD within that taxonomic group. The PD–species curve suggests that even a small number of protected species ("gains"), selected randomly, could deliver a large gain in conserved PD. This is related to the scenario referred to above, where a small number of phylogenetically dispersed species remaining under climate change retained lots of PD. This scenario supports sustainable use—conservation of even a relatively small number of currently useful species could at the same time retain lots of PD and corresponding option values.

However, another scenario demands consideration. While phylogenetically well-dispersed gains can result in higher-PD outcomes (Fig. 3.2, point d), phylogenetically clumped gains may result in lower-PD outcomes (Fig. 3.2, point c). Considering again the Forest et al. (2007) study, the finding that current-use species are phylogenetically clumped suggests that conservation of these species may not represent much conserved PD.

We know that phylogenetically clumped impacts can imply large PD loss. It appears also that conservation of phylogenetically clumped currently used species may not greatly help the overall conservation of PD. A solution to this problem is to somehow integrate the protection of currently used species with conservation that represents the entire phylogenetic tree for that taxonomic group. To examine this, we will explore PD, current uses, and conservation costs in systematic conservation planning.

3.4 PD in Conservation Planning for Sustainable Use

3.4.1 PD and Systematic Conservation Planning

The "phylogenetic sustainable use problem" can be stated as follows: how do we combine conservation of overall PD (and its associated option values) with the conservation of currently valuable species? We will use PD complementarity calculations within systematic conservation planning (SCP) tools to explore this sustainable use problem.

SCP typically is recognized as a family of methods for the efficient selection of areas for the representation and persistence of elements of biodiversity (Sarkar et al. 2006). Most SCP studies are based on biodiversity measures at the species or ecosystems levels. However, phylogenetic diversity measures increasingly are considered in conservation, and methods for incorporating the PD measure into conservation planning continue to be developed. Generally, the goal is to increase the representation of PD when selecting species and/or areas for conservation and management.

The earliest PD studies (Faith 1992a, 1994) linked PD to the cornerstone of SCP, complementarity (see Sect. 3.2.1). These early studies illustrated how PD complementarity values could be used to efficiently select species, or areas, to add to a protected set (see also Faith and Baker 2006). The early software for this phylogenetic SCP was "PD-DIVERSITY" (Walker and Faith 1994), within the DIVERSITY package (Faith and Walker 1993). Other early phylogenetic SCP developments included the integration of costs and probabilities of extinction into PD-based priority setting (Witting and Loeschcke 1995; Weitzman 1998; Faith 2008a). Hartmann and Andre (2013) recently concluded that "using PD in a prioritization process can typically increase biodiversity outcomes by a broad range of 10–220 %".

While PD has been integrated into simple systematic conservation planning algorithms, there are few actual SCP applications (Rodrigues and Gaston 2002; Faith et al. 2004; Sarkar et al. 2006; Strecker et al. 2011). One important development to support practical applications will be methods for taking variable costs of conservation into account. The PD-DIVERSITY software (Faith and Walker 1993; Walker and Faith 1994) implemented PD complementarity for selecting sets of species or areas, but did not enable cost trade-offs (analyses that balance biodiversity and conservation costs).

For our trade-off analyses for this chapter, we adopted another DIVERSITY module, TARGET (for example runs see Faith and Walker 2002). TARGET normally examines species in areas. Here, we shift the input data to features within species. An earlier example of PD systematic conservation planning using this strategy is found in Faith (2008a). The algorithm in this case builds up a list of selected species, by comparing species' PD complementarity values with their weighted conservation costs. A species is added to the conservation set if its PD complementarity value exceeds its weighted cost. A species is deleted from the set during the course of selections if its PD complementarity value becomes less than its weighted cost. Such a deleted species initially (at an earlier stage in adding and deleting species to build up a set) may have yielded a large gain in total net benefit, but addition of other species might have reduced its biodiversity contribution or complementarity value.

The end result of a series of additions and deletions is that the final solution, for any nominated weighting, includes a species if and only if its PD contribution exceeds its weighted cost. The final set minimizes the sum of unrepresented PD and weighted cost. This selection process then is repeated for other nominated weightings, producing a trade-off curve ("efficiency frontier curve") showing alternative solutions. We present examples of this analysis in the next section.

3.4.2 Conservation and Sustainable Use Scenarios

Recent work has illustrated how increases in the magnitude and conservation of estimated ecosystem services can move initial high-biodiversity SCP solutions (sets of conservation priority areas) towards a tipping point in which capacity for regional biodiversity conservation collapses. This problem occurs when the areas offering ecosystem services are all much the same in their regional biodiversity contributions (Faith 2012c). This redundancy in the biodiversity of the ecosystem services areas is analogous to the clumping of evosystem services (currently used species) on a phylogeny. Do increases in the magnitude and conservation of evosystem services analogously move PD-based SCP solutions (sets of conservation priority species) towards a tipping point in which the capacity for PD conservation collapses?

Here, we present one example SCP analysis, for a simple hypothetical phylogenetic tree (Fig. 3.4) and assumed equal ("unit") conservation costs for all species. We varied the assumptions about the extent and phylogenetic distribution of currently used species. In Fig. 3.5a, the black efficiency frontier curve is for the case where there are no currently used species, and the SCP analysis simply maximizes PD for any nominated total cost of conservation.

We next introduced species with current uses. If the current-use value of a species is assumed to imply that the cost of conservation is 0, then there is a clear gain in the net benefits obtained through SCP. For example, suppose that the first 8 members of the large clade (dots; Fig. 3.4) have current use and are selected for conservation action at 0 cost. The green curve (Fig. 3.5a) is the resulting efficiency frontier curve for this case where there is 0 conservation cost for the currently used species. This clearly is a desirable outcome for the sustainable use of biodiversity because a higher level of PD conservation now can be achieved for any given total cost.

However, if there is some unit cost associated with conservation of these currently used species, the trade-off curve changes (Fig. 3.5a). The extent of this shift of the curve towards poor solutions depends, for any given number of current-use species, on the degree to which they are clumped on the phylogenetic tree. For example, suppose all 16 members of the large clade (Fig. 3.4) have current use

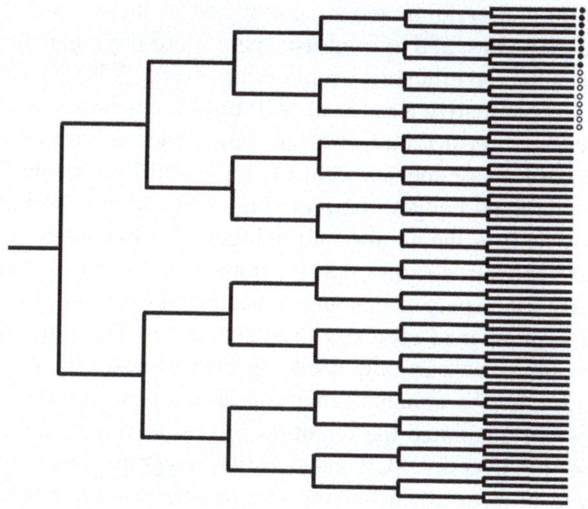

Fig. 3.4 A hypothetical phylogenetic tree with 64 species. Species with dots are those assumed to have current uses in our analyses

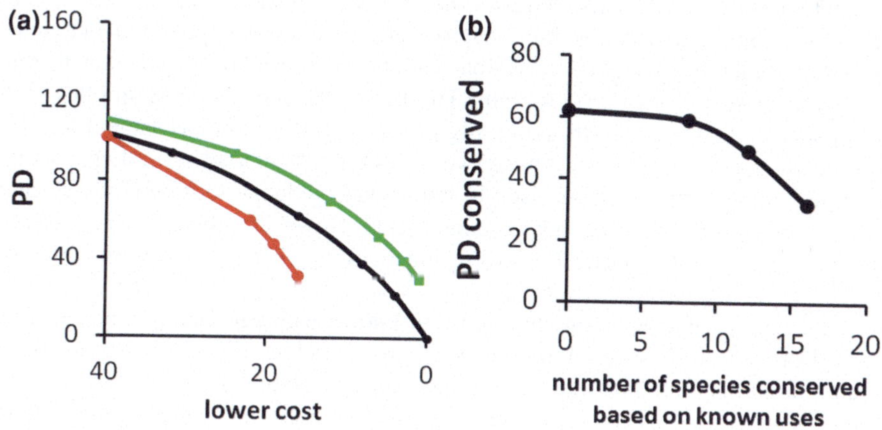

Fig. 3.5 **a** An SCP trade-off space with *vertical axis* equals total PD conserved and *horizontal axis* equals total cost, with lower cost to right. High net benefit solutions are therefore towards the upper right. The *black curve* is for the case where there are no current-use species, and PD is maximized for any nominated cost. The *green curve* is the efficiency frontier curve for the case where there is 0 conservation cost for the current-use species. The *red curve* is the efficiency frontier curve for the case where there is a conservation cost for the current-use species, and these species are phylogenetically clumped. **b** For a fixed budget of 16 units, the plot shows the PD conservation achieved in SCP as the number of clumped current-use species increases. Initially, SCP can find high-PD solutions, but as the number of current-use species increases, the capacity to represent PD drops rapidly

and are selected for conservation action, at unit cost. If there is a conservation cost for the current-use species, and current-use species are phylogenetically clumped, then SCP produces the red efficiency frontier curve (Fig. 3.5a).

In Fig. 3.5b, we summarize a range of SCP results where we maintained a constant total conservation cost ("budget") but varied the number of currently used species. In each case, these species were phylogenetically clumped, as illustrated in Fig. 3.4. For this fixed budget, the plot (Fig. 3.5b) shows the PD conservation level identified by SCP, as the number of phylogenetically clumped, currently used species increases. The curve shows that for our given budget of 16 units, protecting more currently used species means much reduced overall conservation of PD. Initially, for a low number of currently used species, SCP can find high-PD solutions, but as the number of currently used species increases, the capacity to represent PD drops rapidly. We conclude that conservation of currently used species, on its own, does not guarantee the retention of option values that is required for sustainable use of biodiversity. SCP analyses that integrate current uses and option values' goals hold promise for achieving sustainable use, but must be monitored for the kind of tipping point we have described here.

3.5 Discussion

The PD measure reflects expected patterns of feature diversity among species and so provides a way to quantify biodiversity option values. The potential PD gains (or losses) resulting from conservation actions (or impacts) are relevant to the phylogenetic sustainable use problem. The basic PD–species curve implies that initial species losses generally retain high PD, suggesting that occasional loss of current-use species might not reduce overall PD. For example, several species in the legume genus *Pterocarpus*, used in traditional medicine to treat diabetes, are now endangered (Saslis-Lagoudakis et al. 2011). However, other, closely related, species are not endangered. Therefore, much of the PD of the group remains secure.

Conservation of species that have current known uses can maintain overall PD and option values; if currently used species are spread across the phylogeny, they capture more PD than those that are phylogenetically clumped.

On other occasions, current uses may be so phylogenetically clumped that losses can produce tipping points with high-PD loss. This difficulty is raised by the Forest et al. (2007) study's evidence of phylogenetically clumping of currently useful species. Systematic conservation planning that incorporates PD potentially provides a way to overcome this problem; the relatively low PD captured by conservation of currently used species can be complemented by selected conservation of other species. Overall PD conservation then should be high. However, our systematic conservation planning results suggest an important caveat: if there is a conservation budget, conservation of lots of phylogenetically clumped current-use species can use up the budget without much conservation of PD.

Paradoxically, conserving more known-use species can reduce the capacity to conserve PD. Such undesirable sustainability tipping points may be avoided by balancing the conservation of currently valued species and the conservation of overall phylogenetic diversity (PD). In this way, true "sustainable use" preserves not only known uses but also the sustained capacity to find other uses, in other species. We conclude that there is a need to also preserve PD as part of any programme on sustainable use.

Our suggestions fit into a broader picture of the sustainable use of biodiversity. It is now well known that management for current-use species is a major factor in the loss of biodiversity, through associated habitat loss and other factors (e.g. Lenzen et al. 2012). Thus, shared habitat (with the current-use species) is one established factor in considering biodiversity impacts and sustainable use. Here, we have shown that shared evolutionary history (with the current-use species) is another important consideration for sustainable use programmes.

Brazil provides a potential good example of successful phylogenetically based sustainable use programmes. The Biota-FAPESP programme (www.biota.org.br/) has defined a goal for further development of a phylogenetic framework for exploration and assessments, in order to provide a solid basis of sustainable use of the biodiversity. The FAPESP Bioprospecta programme makes effective use of phylogenies in identifying species that potentially have biologically active compounds. The FAPESP Biota programme complements these efforts by also conserving overall phylogenetic diversity in the region—so supporting sustainable use by retaining options for future discovery of useful products in other species (Joly et al. 2010).

By recognizing PD as a measure of option values, and integrating conservation and use, such programmes can capture the core idea (as quoted earlier from Article 2 of the CBD) that sustainable use "does not lead to the long-term decline of biological diversity, thereby maintaining its potential to meet the needs and aspirations of present and future generations".

Acknowledgements DF thanks the Biota programme of the São Paulo Science Foundation (FAPESP) and fellow participants at the international workshop, in São Paulo, Brazil, on "Applied Ecology and Human Dimensions in Conservation Biology". DF thanks participants and fellow members of the scientific advisory board for discussions at the evaluation meeting of the Biota-FAPESP programme, São Carlos, Brazil. Part of this chapter was prepared for a keynote talk at the 49th Annual Meeting of the Association for Tropical Biology and Conservation (ATBC), Bonito, Mato Grosso do Sul, Brazil. DF thanks Lúcia Lohmann and the organizers for support, bioGENESIS members for discussions, and DIVERSITAS for additional funding. LP thanks the National Environmental Research Program (NERP) and the School of Botany at the University of Melbourne for support.

References

Beattie AJ, Hay M, Magnusson B, de Nys R, Smeathers J, Vincent JFV (2011) Ecology and bioprospecting. Aust Ecol 36:341–356

Bordewich M, Semple C (2012) Budgeted nature reserve selection with biodiversity feature loss and arbitrary split systems. J Math Biol 64:69–85

Davies TJ, Buckley LB (2011) Phylogenetic diversity as a window into the evolutionary and biogeographic histories of present-day richness gradients for mammals. Philos Trans R Soc B 366:2414–2425

Díaz S, Hector A, Wardle DA (2009) Biodiversity in forest carbon sequestration initiatives: not just a side benefit. Curr Opin Environ Sustain 1:55–60

Faith DP (1992a) Conservation evaluation and phylogenetic diversity. Biol Conserv 61:1–10

Faith DP (1992b) Systematics and conservation: on predicting the feature diversity of subsets of taxa. Cladistics 8:361–373

Faith DP (1994) Phylogenetic diversity: a general framework for the prediction of feature diversity. In: Forey PL, Humphries CJ, Vane-Wright RI (eds) Systematics and conservation evaluation. Clarendon Press, Oxford, pp 251–268

Faith DP (2008a) Phylogenetic diversity and conservation. In: Carroll SP, Fox CW (eds) Conservation biology: evolution in action, pp 99–115. http://books.google.com.au/books?isbn=0195306783

Faith DP (2008b) Threatened species and the preservation of phylogenetic diversity (PD): assessments based on extinction probabilities and risk analysis. Conserv Biol 22:1461–1470

Faith DP (2010) Biodiversity transcends services. Science 330:1745–1746

Faith DP (2011) Ecosystem services and biodiversity option values. http://www.sciencemag.org/content/330/6012/1745/reply. Cited 10 Feb 2011

Faith DP (2012a) Biodiversity and ecosystem services: similar but different. Bioscience 62:785

Faith DP (2012b) Common ground for biodiversity and ecosystem services: the "partial protection" challenge. F1000 Res. http://f1000research.com/articles/common-ground-for-biodiversity-and-ecosystem-services-the-partial-protection-challenge/

Faith DP (2012c) Tipping points in systematic conservation planning: conservation of ecosystem services may be accompanied by a collapse in regional capacity for biodiversity conservation Abstract P2.47 in: Planet under pressure 2012 conference information

Faith DP (2013) Biodiversity. In: Zalta EN (ed) The stanford encyclopedia of philosophy. url:http://plato.stanford.edu/entries/biodiversity/

Faith DP, Baker A (2006) Phylogenetic diversity (PD) and biodiversity conservation: some bioinformatics challenges. Evol Bioinf 2:121–128

Faith DP, Walker PA (1993) Diversity: a software package for sampling phylogenetic and environmental diversity. Reference and user's guide. v.1.0. In: CSIRO Division of Wildlife and Ecology Canberra, Australia

Faith DP, Walker PA (2002) The role of trade-offs in biodiversity conservation planning: linking local management, regional planning and global conservation efforts. J Biosci 27:393–407

Faith DP, Williams KJ (2006) Phylogenetic diversity and biodiversity conservation. In: McGraw-Hill yearbook of science and technology. McGraw-Hill, New York, pp 233–235

Faith DP, Reid CAM, Hunter J (2004) Integrating phylogenetic diversity, complementarity, and endemism for conservation assessment. Conserv Biol 18:255–261

Faith DP, Magallón S, Hendry AP, Conti E, Yahara T, Donoghue MJ (2010) Evosystem services: an evolutionary perspective on the links between biodiversity and human well-being. Curr Opin Environ Sustain 2:66–74

Forest F, Grenyer R, Rouget M, Davies TJ, Cowling RM, Faith DP, Balmford A, Manning JC, Proches S, van der Bank M, Reeves G, Hedderson TAJ, Savolainen V (2007) Preserving the evolutionary potential of floras in biodiversity hotspots. Nature 445:757–760

Hartmann K, Andre J (2013) Should evolutionary history guide conservation? Biodiv Conserv 22:449–458

Hendry AP, Lohmann LG, Conti E, Cracraft J, Crandall KA, Faith DP, Häuser C, Joly CA, Kogure K, Larigauderie A, Magallón S, Moritz C, Tillier S, Zardoya R, Prieur-Richard A-H, Walther BA, Yahara T, Donoghue MJ (2010) Evolutionary biology in biodiversity science, conservation and policy: a call to action. Evolution 64:1517–1528

Huang S, Davies TJ, Gittleman JL (2012) How global extinctions impact regional biodiversity in mammals. Biol Lett 8:222–225

Isambert B, Bergsten J, Monaghan MT, Andriamizehy H, Ranarilalatiana T, Ratsimbazafy M, Andriniainimanana JR, Vogler AP (2011) Endemism and evolutionary history in conflict over Madagascar's freshwater conservation priorities. Biolog Conserv 144:1902–1909

IUCN (1980) World conservation strategy: living resource conservation for sustainable development. International Union for Conservation of Nature and Natural Resources (IUCN)

Joly CA, Rodrigues RR, Metzger JP, Haddad CFB, Verdade LM, Oliveira MC, Bolzani VS (2010) Biodiversity conservation research, training, and policy in São Paulo. Science 328:1358–1359

Krishnamurthy PK, Francis RA (2012) A critical review on the utility of DNA barcoding in biodiversity conservation. Biodiver Conserv 21:1901–1919

Lenzen M, Moran D, Kanemoto K, Foran B, Lobefaro L, Geschke A (2012) International trade drives biodiversity threats in developing nations. Nature 486:109–112

Mace GM (2003) Preserving the tree of life. Science 300:1707

Mace GM, Wolfgang C, Díaz S, Faith DP, Larigauderie A, Le Prestre P, Palmer M, Perrings C, Scholes RJ, Walpole M, Walther BA, Watson JEM, Mooney HA (2010) Biodiversity targets after 2010. Curr Opin Environ Sustain 2:1–6

McNeely JA (1988) Economics and biological diversity: developing and using economic incentives to conserve biological resources. IUCN, Gland, Switzerland

Millennium Ecosystem Assessment (2005) Ecosystems and human well-being: biodiversity synthesis. World Resources Institute, Washington DC

Morlon H, Schwilk DW, Bryant JA, Marquet PA, Rebelo AG, Tauss C, Bohannan BJM, Green JL (2011) Spatial patterns in phylogenetic diversity. Ecol Lett 14:141–149

Nipperess DA, Faith DP, Barton K (2010) Resemblance in phylogenetic diversity among ecological assemblages. J Veg Sci 21:1–12

Pacharawongsakda E, Yokwai S, Ingsriswang S (2009) Potential natural product discovery from microbes through a diversity-guided computational framework. Appl Microbiol Biotechnol 82:579–586

Penafiel D, Lachat C, Espinel R, Van Damme P, Kolsteren P (2011) A systematic review on the contributions of edible plant and animal biodiversity to human diets. EcoHealth 8:381–399

Perrings C, Naeem S, Ahrestani F, Bunker DE, Burkill P, Canziani G, Elmqvist T, Ferrati R, Fuhrman J, Jaksic F, Kawabata Z, Kinzig A, Mace GM, Milano F, Mooney H, Prieur-Richard AH, Tschirhart J, Weisser W (2010) Conservation. Ecosystem services for 2020. Science 330:323–324

Redford KH, Richter BD (1999) Conservation of biodiversity in a world of use. Conserv Biol 13:1246–1256

Reid WV, Miller KR (1989) Keeping options alive: the scientific basis for conserving biological diversity. World Resources Institute, Washington DC

Rodrigues ASL, Gaston K (2002) Maximising phylogenetic diversity in the selection of networks of conservation areas. Biol Conserv 105:103–111

Sarkar S, Pressey RL, Faith DP, Margules CR, Fuller T, Stoms DM, Moffett A, Wilson KA, Williams KJ, Williams PH, Andelman S (2006) Biodiversity conservation planning tools: present status and challenges for the future. Ann Rev Environ Resour 31:123–159

Saslis-Lagoudakis CH, Klitgaard BB, Forest F, Francis L, Savolainen V, Williamson EM, Hawkins JA (2011) The use of phylogeny to interpret cross-cultural patterns in plant use and guide medicinal plant discovery: an example from *Pterocarpus* (Leguminosae). PLoS ONE 6:e22275. doi:10.1371/journal.pone.0022275

Smith MA, Fisher BL (2009) Invasions, DNA barcodes, and rapid biodiversity assessment using ants of Mauritius. Front Zool 6:31

Smith WL, Wheeler WC (2006) Venom evolution widespread in fishes: a phylogenetic road map for the bioprospecting of piscine venoms. J Hered 97:206–217

Strecker AL, Olden JD, Whittier JB, Paukert CP (2011) Defining conservation priorities for freshwater fishes according to taxonomic, functional, and phylogenetic diversity. Ecol Appl 21:3002–3013

Thuiller W, Lavergne S, Roquet C, Boulangeat I, Lafourcade B, Araujo MB (2011) Consequences of climate change on the tree of life in Europe. Nature 470:531–534

Tulp M, Bohlin L (2002) Functional versus chemical diversity: is biodiversity important for drug discovery? Trends Pharmacol Sci 23:225–231

UNCSD (2012) The future we want. In: United Nations conference on sustainable development Rio de Janeiro, Brazil. http://www.uncsd2012.org/content/documents/727The%20Future%20We%20Want%2019%20June%201230%20pm.pdf. Cited 1 Aug 2012

Walker PA, Faith DP (1994) DIVERSITY-PD: procedures for conservation evaluation based on phylogenetic diversity. Biodiv Lett 2:132–139
Weitzman ML (1998) The Noah's ark problem. Econometrica 66:1279–1298
Witting L, Loeschcke V (1995) The optimization of biodiversity conservation. Biol Conserv 71:205–207
Yesson C, Culham A (2006) A phyloclimatic study of Cyclamen. BMC Evol Biol 6:72

Chapter 4
Adaptation and Evolution in Changing Environments

Luís Miguel Rosalino, Luciano M. Verdade and Maria Carolina Lyra-Jorge

Abstract In a planet in continuous change, where humans and their activities have enhanced this modification's rate, wildlife had to develop mechanisms to cope with this change to avoid extinction. The challenges imposed by agricultural landscapes, often associated with production cycles and intensification, has led to the extinction of many populations or species unable to track resources or change their behavioral or biological processes to allow them to use the available resources. However, not always the story has a negative outcome. In the present chapter, we present several examples of acclimation and adaptation processes of species trying to survive in changing environments, which include morphological, physiological, and behavioral adaptations (i.e., with genetic implications) and behavioral acclimations. However, there is a threshold of changes that a species may manage to overcome. This limit is species specific and directly related to the species natural history, plasticity and genetic structure and diversity.

L. M. Rosalino (✉)
Centro de Biologia Ambiental—Faculdade de Ciências de Lisboa,
Universidade de Lisboa, Ed. 2, Campo Grande, 1749-016 Lisboa, Portugal
e-mail: lmrosalino@fc.ul.pt

L. M. Rosalino
Laboratório de Ecologia Isotópica, CENA/Universidade de São Paulo,
Caixa Postal 96, Piracicaba, SP 13416-000, Brazil

L. M. Verdade
Centro de Energia Nuclear na Agricultura, Universidade de São Paulo,
Caixa Postal 96, Piracicaba, SP 13416-000, Brazil
e-mail: lmverdade@usp.br

M. C. Lyra-Jorge
Curso de Ciências Biológicas, Universidade de Santo Amaro,
Rua Prof. Eneas de Siqueira Neto, 340, São Paulo, SP 04829-300, Brazil
e-mail: carolina_ecologia@hotmail.com

L. M. Verdade et al. (eds.), *Applied Ecology and Human Dimensions in Biological Conservation*, DOI: 10.1007/978-3-642-54751-5_4,

4.1 Wildlife Challenges in Changing Environments

Throughout the history of the planet Earth, life has been shaped by the interaction between abiotic or physical (e.g., patterns of water and air circulation, movement of tectonic plates, solar radiation) and biotic (e.g., inter- and intraspecific relations) factors. Although changes have been occurring since life exists, conservation biology has been mostly concerned about those anthropic changes (i.e., directly related to humans' presence or their activities) that may affect biodiversity in general and species and/or population in particular, inducing habitat destruction, species extinction, and populations' decline (Caro 2007). To achieve their goals, conservation biologists aim to identify the threatening processes that cause detrimental effects on the population and species survival, abundance, distribution, and evolution (Lindenmayer and Burgman 2005). The most commonly recognized anthropic threatening processes are habitat destruction (Travis 2003), competition with invasive species (e.g., Yamada and Sugimura 2004), change in interspecific relations (e.g., Colwell et al. 2012), overexploitation (e.g., Bodmer et al. 1997), pollution (Mann et al. 2009), and dispersal of diseases (e.g., Briones et al. 2000). However, these processes did not arise simultaneously as our species was differentiating itself from other primates in African plains. Most of them are intimately related to the human technological development (e.g., agrochemicals), as it is their impact on biodiversity. Even so, some of the above-mentioned anthropic shapers have been acting upon the landscape, and species, for many millennia.

Man-shaped environments are becoming dominant in many parts of the world, and their characteristics and impacts on wildlife vary according to the activities that have molded these environments. Large dam and small hydroelectric power plants transformed many of the world's lotic and pristine rivers into lentic environments, where invasive species are common (often introduced for sport fishing; Collares-Pereira et al. 2000) and sediments accumulate. Mining (underground or open pit) destroyed many mountainous areas often transforming mountain environments into plains or even lakes, polluting adjacent rivers or streams as the water emerging from the debris may contain toxic compounds (e.g., Wayland and Crosley 2006). Oil and gas extraction (including the associated power lines, roads, and collection stations) has also been responsible for the destruction and pollution of many pristine environments (see Holdway 2002 for a review of the effects on temperate and tropical marine ecological processes).

However, one of the most impacting activities that have altered the landscapes, transforming drastically the world, is agriculture *lato sensu* (i.e., including cultivation of the soil for the growth of crops—agriculture *sensus strictus*—besides cattle raising and silviculture). The use of fire by Palaeolithic hunter-gatherers, at least 200,000 years ago, significantly altered wildlife habitats (Naveh 1975). However, the implementation of agricultural practices for crop production 10,000–12,000 years ago dramatically changed the face of the Earth (Blondel 2006). This activity was associated with settlements and the radical transformation of a hunter-gatherer way of life into a productive economy, generator of surplus, and trade of commodities based on land exploitation. Since then, it has spread globally, affecting all worlds' environments, creating agricultural landscapes, which include areas with agricultural production patches/matrix (agroecosystems; e.g., cereal fields), upon which native patches are embedded (e.g., cork oak woodlands in the Mediterranean regions). So agriculture has become the major activity responsible for the conversion of pristine areas.

These complex landscapes have introduced a different kind of heterogeneity that wildlife must cope with in order to survive. The pristine or the so-called natural environments present a spatial heterogeneity derived from the different structures of the matrix and of all the different patches corresponding to the puzzle pieces composing the landscape (e.g., riparian vegetation within a forest). We must also consider that to this two-dimensional heterogeneity, we should add a third axis, composed by a different vertical level that might include a ground/herbaceous, a scrub, and/or a canopy level. All these create different niches that might be used by various species, allowing them to co-exist (e.g., Rosalino et al. 2011). However, this variation also imposes some limitation on how they use the available resources and species have evolved to cope with it.

Agricultural landscapes changed radically the scenario, as the typical spatial heterogeneity of pristine environments was replaced by a more homogeneous spatial pattern associated with intensive production (e.g., sugarcane plantation, cereal fields, *Eucalyptus* plantations). In addition, in many of such environments, the vertical component of the habitat is also lost (e.g., soybean plantations). Inversely, agricultural landscapes are characterized by a relatively high temporal heterogeneity in the habitat structures intimately associated with the production cycles. For example, in 12–18 months, areas devoted to sugarcane production evolve from bare soil to densely vegetated areas, encompassing 3–6 m high plants (depending on the variety used and area of cultivation), corresponding to 100 ton ha^{-1} of green matter (FAO 2012). On the other hand, *Eucalyptus* plantations have an extended successional change, as harvesting can occur after 6–11 years of plantation, reaching an arboreal stratum usually with few understory (Silva 2007). In other less intensive productions, as in Mediterranean traditional multifunctional landscapes, where crop productions are associated with forestry (cork extraction) and cattle raising, the change in the landscape is less radical as the forest persists for decades (with an yearly intervention for cork extraction), cattle grazes throughout the year, and multispecies crops associated with orchards are permanently managed (Pinto Correia and Vos 2004). For wildlife, these temporal changes can act as simple fluctuations associated with fast production cycles if we consider animal species with longer generation times, successional changes in longer production cycles for short-generation species, or evolution drivers if extended for a long period of time (e.g., many decades or centuries) (Preston 1960).

In the face of these alterations, species have several pathways they can follow. If they are unable to acclimate or adapt to the changes, they will get extinct. If they are mobile enough, they can survive by tracking their favored resource or habitat. Finally, they may be flexible enough to change to cope with the environmental alteration, due to polymorphism (alternative alleles that will enhance the fitness in the new environments), phenotypic plasticity (distinct phenotypes may allow survival in specific environmental conditions), adaptive tolerance (physiological or behavioral flexibility to changes), or adaptive versatility (a particular structure or behavior may increase fitness in new environments) (Potts 2004).

4.2 The Extreme Outcome: Extinction and Biodiversity Loss

Human activities may have irreversible impacts on biodiversity, expressed as gene, populations, and species losses, leading to the extinction of those unable to acclimate or adapt in time to the occurred changes. If to the transformation of the environments by man, species do not respond fast enough (or move to track their crucial resources), and as a consequence, their population sizes diminish abruptly, there is a loss of genetic information. This loss associated with bottlenecks and inbreeding, with a consequent decrease in the individual fitness to cope with changes occurring in the environments (e.g., change in available resources, diseases), may cause extinctions (Lindenmayer and Burgman 2005).

Several paths may lead to this population decreases, and the molding factors may act independently or synergistically: habitat and diet specialism, low reproductive output, large-scale habitat destruction, pollution, and restricted distribution range. As mentioned before, some species may be too specialized in exploring particular resources (i.e., preference for a restrict subset of resources independent of its abundance) that lack of adaptive tolerance to face environmental changes. In a review study, Colles et al. (2009) showed several examples of a correlation between habitat specialization and increased extinction risk in different taxa: birds, bats, bumblebees, and plants. For example, Safi and Kerth (2004) showed that bats specialized in foraging in closed habitats such as forests or riparian areas face higher extinction rate due to deforestation. A similar pattern was detected by Owens and Bennett (2000) in a meta-analysis to assess the ecological basis of birds' extinctions, which found that when the probability of extinction was intimately related to habitat loss, there was a clear association between the species more prone to be extinct and those characterized as habitat specialists. On the other hand, this association is often related to the fact that habitat specialists explore ecological niches frequently limited to restricted areas. This constrains the species distribution range, and thus, any environmental change that may affect resources availability in that particular region might have drastic effect on the species survival.

The specialization on the use of other resources follows a similar pattern. Food resources availability changes and diet specialisms can synergistically contribute to species extinction. For example, species that rely on a particular prey may not survive if the prey disappears or strongly reduces its population sizes. Two well-described cases are the dependence of the Iberian lynx (*Lynx pardinus*), the world's most threaten felid (Palomo and Gisbert 2002), and the Spanish imperial eagle (*Aquila adalberti*) on the wild rabbit (*Oryctolagus cuniculus*), their main prey. Lynxes and imperial eagles have evolved predatory behaviors adapted to prey rabbits. However, rabbits' populations have crashed in most of the areas of sympatry with both predators due to climatic changes, emergence of rabbit diseases, and shifts in land use (Real et al. 2009; Ferrer and Negro 2004). This decline in their main prey abundance was one of the reasons pointed out for lynx and imperial eagle's critical current population status, although other preys are still available. In the high-arctic Greenland, Schmidt et al. (2012) also detected that after a lemming collapse (a rodent which is the base of a trophic guild in those northern landscapes), snowy owls (*Bubo scandiacus*) ceased breeding and the fledglings' production decreased by more than 98 %, threatening the species survival.

Large-scale habitat destruction is one of the major factors that enhance the extinction risk of many species and/or group of species. Even if the species are not habitat specialists and have dispersal abilities, the large-scale character of this anthropic change may drive species to extinction. For example, Australian marsupials' extinction risk is highly correlated with the proportion of overlap with the range of introduced species (namely sheep), and the mechanism driving this pattern is probably habitat degradation associated with pastoral expansion (Fisher et al. 2003). Also in south-central New South Wales, Australia, in an area where the native vegetation was extensively cleared and transformed into an agriculture matrix, the reptiles disappeared from 90 % of the studied areas, being restricted to remnant vegetation patches (Driscoll 2004). The species that presented higher decline were those that showed a lower dispersal capacity, i.e., they could not migrate to track resources. Even species that may have some capability to move in search of alternative resources might not have success, since this could be a challenging task as the scale of habitat conversion may also jeopardize the ecological corridors across the less favorable landscape (Colles et al. 2009).

The intensification of agricultural practices added extra-challenges for species trying to live in such environments, some of which are lost battles. The best example is the increased use and toxicity of many agrochemicals, most of which we cannot be sure of all the actions and impacts they have on the environments and wildlife (e.g., due to bioaccumulation). Some of these chemicals are transported to underground and above-ground water sheds affecting mainly species that depend on these riparian environments, such as amphibians. Mann et al. (2009) identified several consequences to amphibians of the exposition to agrochemicals such as teratogenesis and abnormal sexual development, endocrine disruption, precocious or delayed metamorphosis, external malformations, among others. All these alterations affect the fitness and may lead to local extinction.

Finally, the reproduction strategy adopted by the different species may also be responsible for enhancing the extinction risk in altered landscapes. When a species has low reproductive output, due to reproductive strategies that include few offspring, single yearly reproductive season, large parental investment, and/or later sexual maturity, it may have difficulties in adapting to changing environment, declining rapidly (Kupfer and Franklin 2009; Lomolino and Perault 2007). When the population of such species decline below a threshold, they will have huge difficulties in recovering, being thus susceptible to local or global extinctions. A good example of such difference in reproductive strategies and its impact is described by King and Moors (1979) to explain the extinction of polecats (*Mustela putorius*) and pine marten (*Martes martes*) in England in the nineteenth century. These mustelids were equally persecuted by gamekeepers, together with weasels (*Mustela nivalis*) and stoats (*Mustela erminea*), but contrary to the two later species (whose reproductive strategy involves early sexual maturity, short period of parental care, and generation time), they did not manage to recover their number. Nevertheless, authors recognize that other factors may also be concurrently involved (e.g., deforestation). In New Zealand, the survival of the takahe (*Notornis mantelli*), a bird species with long life expectancy and low annual productivity of offspring, is dependent on the low adult mortality. If this population parameter increases (due to, for example, invasive species or habitat destruction), the species may become threatened.

4.3 The Survivors' Option: Acclimation and Adaptation

Fortunately, the challenges imposed by the anthropic changes in the landscape are not a curse that condemns all wildlife to extinction. Many species manage to acclimate or adapt to new conditions. Acclimation can be considered as reversible physiological, behavioral, or morphological alterations that enable an organism to tolerate changing environmental conditions, whereas adaptation is a genetic change in characteristics that enhance organisms' fitness in changing environments, by allowing differential survival and reproduction (Chevin et al. 2010). Acclimation, therefore, occurs at the individual level within a single generation, whereas adaptation implies alleles frequency change at the population level along generations.

Some species have achieved to get advantages of the surplus of food and refuges available in anthropic environments. These species that, although continuing wild, live in close contact with man may be named domestic, which should not be confused with those called domesticated species (Descola 1996). The former are those that live inside our cities or houses as cockroaches (insects Blattidae), house sparrows (e.g., *Passer domesticus*), pigeons (e.g., *Columba* spp.), blackbird (*Turdus merula*), mice (e.g., *Mus musculus*), rats (e.g., *Rattus norvegicus*), rabbits (e.g., *Oryctolagus* spp.), or even carnivores (e.g., red fox, *Vulpes vulpes*, stone marten, *Martes foina*, and badger, *Meles meles*) (Luniak 2004). The later were molded by man for particular purposes (e.g., meat and milk production, protection), being incorporated into humans' social structure, becoming mostly isolated from wild species, with its breeding and food supply being perdominantly under human control (Clutton-Brock 1992). However, this ability of domestic species is not unlimited. For example, phenotypic plasticity, one of the mechanisms allowing survival in changing environments, is limited by constitutive cost, associated with the maintenance of the physiological machinery responsible for the plasticity, and by the induced costs, related to the physiological cost dependent on the degree of change (Chevin et al. 2010). Furthermore, evolution/adaptation also affects the interaction structure of natural communities that limit the success of the process (Thompson 1998).

Considering only the wild species that use altered landscapes and the domestic ones, we reviewed some examples of acclimation and adaptation of invertebrate and vertebrate species to changing environments. These data are presented in

Table 4.1, where we describe the species involved, the environmental change, the trait observed, the type of acclimation/adaptation, and the reference. The review is not intended to be a thorough review of the literature on this subject, but only to present different examples of how species might be able to handle anthropic disturbance and landscape change. Previous reviews about similar subjects showed the relatively high capacity of vertebrates to hunting pressure (Verdade 1996) and of mesocarnivores (i.e., medium-sized Carnivora) to land use change related to agriculture (Verdade et al. 2011).

The quickest and simplest way to cope with changing environments is to use the ecological and physiological plasticity derived from the genetic structure of each species and acclimate. This process is frequent throughout most of the taxonomic groups (see Table 4.1). One of the most common acclimation processes is the use of new or recently abundant food resources. Many species manage to change their food habits and take advantage of many trophic resources directly (e.g., fruit groves) or indirectly (e.g., predation upon rodents that survive using human agriculture productions) related to humans (Table 4.1).

Starlings (*Sturnus vulgaris*) are a common European omnivorous bird that may reach high densities, especially in agricultural areas. Feare and Wadsworth (1981) showed that these passerines have developed the capacity to consume cattle food (e.g., barley) that is usually disposed in yards or in open buildings. These authors estimate that in less than 3 months, in one farm, starlings were responsible for the loss of 9 % (approximately 12 ton) of 132 ton of cattle food used in the farm.

In the Neotropics, capybaras (*Hydrochoerus hydrochaeris*) living in an area where the semi-deciduous subtropical forest was replaced by pastures and sugarcane plantations use the former landscape unit as shelter and sugarcane plantations as foraging ground (Ferraz et al. 2007). The species can dramatically increase its population density in agricultural landscapes because of its capacity to eat domestic C4 plants (Verdade and Ferraz 2006).

Even carnivores manage to shift their diet and consume human-related foods. Some of these predator species living in Mediterranean Europe specialized in using such resources, even with an associated high mortality risk. In southwestern Portugal, Eurasian badgers (*M. meles*) are considered seasonally specialized in the consumption of olives (*Olea europaea*), which are cultivated by man to produce oil (Rosalino et al. 2005). On the other hand, other populations of these carnivores, living in areas where olive yards are scarce or inexistent, consume mainly earthworms (Goszczynski et al. 2000). Otters (*Lutra lutra*) and wolves (*Canus lupus*), on the other hand, started using resources that are economically important to man, and this acclimation originated some conflicts. With the increasing market demands for fish in Europe, fish farms were considered a valuable and sustainable enterprise. Otter found in fish farms an alternative food resource and started using these structures as foraging grounds. In such circumstance, stoked fishes constituted the bulk of its diet, reaching 87 % of the total consumed biomass (Marques et al. 2007). This obviously has raised some conflicts with fish farmers that may threaten otters' conservation.

Table 4.1 Examples of animal species acclimation/adaptation to changing environments

Species	Environmental change	Alteration traits	Acclimation/adaptation	Reference
Limnodrilus hoffmeisteri (Annelida, Oligochaete)	Metal pollution (e.g., Cd, Ni, and Co)—wastewaters from battery factory	Increased survival	Resistance genetically determined (adaptation)	Klerks and Levinton (1989)
Arion subfuscus and *Deroceras reticulaturn* (Mollusca)	Metal pollution (Pb, Cd and Zn)—Pb/Zn mine	Restricted tissue accumulation	Phenotypic-driven reduction in metal accumulation (acclimation)	Greville and Morgan (1991)
Porcellio scaber (Arthropoda, Isopoda)	Metal pollution (Cd, Cu or Zn)—lead mine site	Development of detoxification mechanisms (deposition of cadmium in intracellular granules or the formation of organic complexes)	Genetic adaptation	Donker and Bogert (1991)
Euphydryas editha (Arthropoda, Insecta)	Introduced plant by cattle ranchers	Diet change	Genetic change in preferences (adaptation)	Singer et al. (1993)
Jadera haematoloma (Arthropoda, Insecta)	Introduced plant	Change in egg size, juvenile survivorship, and fecundity	Adaptive responses to host nutritional differences (adaptation)	Carroll et al. (1998)
Jadera haematoloma (Arthropoda, Insecta)	Introduced plant	Morphological change	Genetic adaptation	Carroll and Boyd (1992)
Prodoxus quinquepunctellus (Arthropoda, Insecta)	Introduced plant	Emergence and ovipositor characters' alteration	Genetic adaptation	Groman and Pellmyr (2000)
Culex pipiens (Arthropoda, Insecta)	Presence of anthropic structures (underground tunnels)	Reduction or suppression of diapause	Genetic adaptation	Byrne and Nichols (1999)
Poecilia reticulata (Vertebrata, Fish)	Commercial fishing/higher predation	Change in reproduction strategies	Genetic adaptation	Reznick and Ghalambor (2005)
Oncorhynchus nerka (Vertebrata, Fish)	Translocation of individuals to new habitats	Morphological change	Morphological adaptation	Hendry and Quinn (1997)

(continued)

Table 4.1 continued

Species	Environmental change	Alteration traits	Acclimation/adaptation	Reference
Rana arvalis (Vertebrata, Amphibia)	Anthropogenic acidification of lakes	Increased survival and diminished cost of development and growth	Genetic adaptation	Räsänen et al. (2003)
Rana aurora (Vertebrata, Amphibia)	Introduction of an invasive predator	Reduction in movement and increased use of shelter	Behavioral adaptation	Kiesecker and Blaustein (1997)
Amphibians guild (Vertebrata, Amphibia)	Road presence	Movement pattern change	Behavioral acclimation	Gravel et al. (2012)
Pseudechis porphyriacus and *Dendrelaphis punctulatus* (Vertebrata, Reptilia)	Introduction of toxic invasive cane toads (*Bufo marinus*)	Morphological change	Morphological adaptation	Phillips and Shine (2004)
Snakes (Reptile) (Vertebrata, Reptilia)	Introduction into an anthropic environment (city)	Use of resources of anthropic origin	Behavioral acclimation	Eterovic and Duarte (2002)
Carpodacus mexicanus (Vertebrata, Aves)	Introduction into new environments	Development of latitudinal migration	Behavioral adaptation	Able and Belthoff (1998)
Falco naumanni (Vertebrata, Aves)	Scarcity/destruction of natural nest sites and abundance of anthropic structures	Nesting in old buildings	Behavioral acclimation	Negro and Hiraldo (1993)
Sturnus vulgaris (Vertebrata, Aves)	Introduction of artificial feeding for cattle	Use of alternative food resources	Behavioral acclimation	Feare and Wadsworth (1981)
Peromyscus leucopus (Mammalia, Rodentia)	Landscape change due to agriculture	Change in landscape use and movement	Behavioral acclimation	Wegner and Merriam (1990)
Praomys tullbergi and *Lophuromys* spp. (Mammalia, Rodentia)	Landscape change due to agriculture	Change in landscape use	Behavioral acclimation	Barnett et al. (2000)
Hydrochoerus hydrochaeris (Mammalia, Rodentia)	Landscape change due to agriculture	Change in landscape use and diet	Behavioral acclimation	Ferraz et al. (2007)
Oryctolagus cuniculus (Mammalia, Lagomorpha)	Introduction into new environments	Morphological changes	Morphological adaptation	Williams and Moore (1989)

(continued)

Table 4.1 (continued)

Species	Environmental change	Alteration traits	Acclimation/adaptation	Reference
Meles meles (Vertebrata, Carnivora)	Introduction of agriculture patches	Diet change	Behavioral acclimation	Rosalino et al. (2005)
Canis latrans (Vertebrata, Carnivora)	Introduction of anthropic foods	Change in diet and densities	Behavioral acclimation	Fedriani et al. (2001)
Canis lupus (Vertebrata, Carnivora)	Introduction of livestock herds	Diet change	Behavioral acclimation	Vos (2000)
Martes foina (Vertebrata, Carnivora)	Presence of cities	Use of anthropic hideouts	Behavioral acclimation	Tóth et al. (2009)
Vulpes vulpes (Vertebrata, Carnivora)	Increase in food availability in agriculture areas	Morphological change	Morphological adaptation	Yom-Tov et al. (2007)
Lutra lutra (Vertebrata, Carnivora)	Presence of fish farms	Diet change	Behavioral acclimation	Marques et al. (2007)
Elephas maximus (Vertebrata, Proboscidea)	Presence of an agriculture land use mosaic	Behavioral change	Behavioral acclimation	Kumar and Singh (2011)

Wolves have also started using domesticated prey as their wild prey became scarce. For example, in Western Iberian, wolf diet was almost exclusively composed by livestock, especially goats (Vos 2000). Wolves attacks on large flock of >100 heads have also resulted in conflicts with an increasing predator's persecution by man. Similar conflicts have been described in the Neotropics between pumas (*Puma concolor*) and livestock producers (Verdade and Campos 2004; Palmeira et al. 2008).

Even in urban environments, wild species manage to take advantage of human detritus or the species that are abundant due to those resources (e.g., rodents). Eterovic and Duarte (2002) found at least seventy-six individuals of sixteen exotic snake species living in São Paulo City (São Paulo State, Brazil), and although no successful colonization was confirmed (i.e., reproduction), specimens survived after release in an alien environment by taking advantage of the huge rodents' populations associated with human detritus.

In Santa Monica Mountains of California (USA), coyotes (*Canis latrans*) have a highly variable diet, but the population living in the southern Cheeseboro and Palo Comado Canyons, a highly urbanized area, consume a high proportion of anthropic food (e.g., trash, livestock, and domestic fruit), reaching 25 % of their prey items (Fedriani et al. 2001). These authors believe that the use of the highly abundant anthropic food is also responsible for the higher densities of this canid registered in that area. Similarly, feral dogs and cats became the major predators of local wild mammals on a suburban area in southeastern Brazil (Gheler-Costa et al. 2002; Campos et al. 2007).

However, the acclimation to changing environments is not only expressed as diet changes, allowing species to survive even in drastically altered environments. For example, *Arion subfuscus* and *Deroceras reticulaturn*, two slug species (mollusks), living in contaminated Pb/Zn mining areas show some tolerance to metal accumulation due to certain phenotypic alteration like secreting extracellular metal chelators that reduce trans-epithelial metal transport and by increasing the number of metal-sequestering ligands in the tissues which reduce the concentrations of free metal ions (Greville and Morgan 1991). Even freshwater turtles appear to strive in highly contaminated water courses in southeastern Brazil (Piña et al. 2009).

Other organisms changed their movement and land use pattern to cope with changes. Roads are one of the main amphibian mortality causes, especially if they cross areas used by animals to migrate between, for example, natal ponds and other terrestrial habitat. Road's presence also alters individual movement behavior, due mainly to disturbances associated with this linear structure such as light, noise or vibration, or their open canopy structure which might deter amphibians from crossing (Gravel et al. 2012).

The presence of landscape patches with high food availability may also function as attractants and lead animals to shape their movement patterns according to those patches' locations. On agricultural landscapes in Ontario, Canada, *Peromyscus leucopus* mice molded their movement and land use pattern in relation to the location of crop patches, which were recently introduced, to track food produced by agriculture (e.g., corn, barley) (Wegner and Merriam 1990).

In cocoa groves of Sierra Leone, rodents (e.g., *Praomys tullbergi* and *Lophuromys* spp.) have acclimate to use these agriculture patches, using them intensely, as they provide refuges for forested species in a highly deforested area, encompassing shading and brushy microhabitats that those species use (Barnett et al. 2000). Some species of small rodents are more abundant in sugarcane fields than on their original habitats (Gheler-Costa et al. 2012), probably due to a high food availability in those habitats. Others become residents at *Eucalyptus* plantations (Martin et al. 2012) apparently showing subtle morphological changes even after a small period of land use change (Rosalino et al. 2013).

Other species have adjusted their group behavior in order to survive in altered landscapes. In the Anamalai Mountains, India, Asian elephants (*Elephas maximus*) have changed their vigilance behavior to cope with the high disturbance in tea plantations, by increasing their alertness and decreasing the interindividual distances. These alterations reduce the probability of being detected or surprised by humans when crossing these patches (Kumar and Singh 2011).

As for acclimation processes in altered environments that evolve food habits change, other species have managed to develop ecological strategies that enable them to use anthropic refuge resources, inside cities. The lesser kestrel (*Falco naumanni*) and the stone marten (*Martes foina*) are two of those species. Krestels are common visitors and breeders in southern Spanish churches, where they find suitable refuge and nest sites in holes located high above the ground to reduce predation risks (e.g., by domestic cats and rats) (Negro and Hiraldo 1993). These nests are highly conspicuous and easily identified. However, the use of the city space by some carnivores is not commonly perceived by the population, due to their elusive and nocturnal character. For example, in Budapest (Hungary), stone martens use attics, roof spaces, suspended ceilings, and church towers as safe refuges (Tóth et al. 2009). They also use old houses with courtyards, small gardens, and circular galleries where they can also find food (e.g., fruits and mice).

When changes in the environment are continuous and long-lasting and/or the species involved have short generation times, these acclimation processes may lead to adaptation, with the consequent alteration of the genetic structure that allows the improvement in the organisms' fitness in changing environments. This evolution might be rapid (e.g., 13 generations in introduced sockeye salmon *Oncorhynchus nerka*—Hendry et al. 2000) or take more than one hundred generations (e.g., *L. pardinus*—Pertoldi et al. 2006). Whatever the time/generation scale involved, adaptation may be driven by coevolution or alteration of interspecific interactions, introduction of exotic/non-native species, or major environmental change (e.g., biophysical alteration of habitats) (Thompson 1998). In Table 4.1, we highlighted some of these examples.

The most perceptible surrogate of the occurrence of an adaptive process is the alteration of species morphological traits. For example, in Florida (USA), the goldenrain tree (*Koelreuteria elegans*) was introduced from Asia, in the 1950s, and the soapberry bugs (*Jadera haematoloma*) started to use this plant as host species instead of the native perennial balloon vine (*Cardiospermum corundum*). Carroll et al. (1998) notice an adaptive response to the introduced host nutritional differences. Those bugs adapted to the new host expressed a consistent change in egg size (which became smaller), juvenile survivorship (higher on the introduced host), and fecundity (lower on native and higher on introduced host), which enabled this bugs to enhance their fitness in the presence of the introduced host. This bug species uses its beak to reach seeds located inside the plant fruits, and its lengths are related to fruit size. In the presence of smaller non-native species fruits, the soapberry bugs in Florida showed a drastic reduction in beak lengths, consistent along generations (Carroll and Boyd 1992).

Another phytophagous insect (*Prodoxus quinquepunctellus*) also showed some adaptation to introduced plants (*Yucca* spp.). This Lepidoptera oviposits its eggs into yucca inflorescence, during the flowering period. By comparing two *Prodoxus* populations, from different areas, which used two yucca species (one native and one introduced, with different flowering periods) for laying eggs, Groman and Pellmyr (2000) detected that moth emergence pattern was correlated with the host plants' flowering and that ovipositor size and structure were adapted to the morphology of each plant inflorescence. These differences were corroborated by the analysis of the genetic structure of the moth populations feeding and reproducing on the two different *Yucca* species, which was considerably different (Groman and Pellmyr 2000).

Adaptive processes involving morphological alteration in vertebrates are also common in most groups. Sockeye salmons (*Oncorhynchus nerka*) may spawn in streams and on beach, and their body morphology often reflects this difference, with beach spawning population presenting deeper bodies, which have deleterious effect in fast-flowing streams due to less efficient shape hydrodynamics. Lake Washington (Washington State, in USA) salmon populations were mostly composed by transplanted individuals with the same origin (and the same spawning habitat). Nowadays, they use different areas to spawn in the lake, which is reflected on their body shape (e.g., deeper bodies in beach spawning individual), a character variation that has a genetic basis. Although the origin populations of the salmons in Lake Washington are the same, individuals are adapted to different conditions of the lake where they were introduced (Hendry and Quinn 1997).

Other species have shaped their body morphology to cope with toxic introduced preys. The cane toad (*Bufo marinus*) was introduced in Australia in 1935 having extended its range since then. As a novel species, it became the prey of many Australian predators, especially snakes. However, its high toxicity has imposed a survival stress to predators that had to adapt to cope with this alien species. The snake prey size is limited by their gape size, and consequently head size, and smaller individuals have relatively larger heads. Thus, a serpent with a smaller gape will have less probability in ingesting a toad large enough to seriously poison and kill it. Such process was detected in *Pseudechis porphyriacus* and *Dendrelaphis punctulatus* snakes, which presented a reduction in gape size and an increase in body length in the presence of the toxic cane toad (Phillips and Shine 2004).

Also in Australia, other examples can be found, such as the morphological alteration, with a genetic basis, of the introduced European wild rabbit (*O. cuniculus*).

This species is adapted to the extreme temperature of the Australian continent (i.e., optimizing the thermoregulation) by developing a leaner body core and longer ears (Williams and Moore 1989). In an area of the rabbit's native ranges (Iberia), one of its known predators has also changed its body size as a response to the high availability of human-related food resources. Foxes' skull analysis has shown that during the twentieth century, animals from agricultural areas had significantly larger skulls (and consequently larger bodies) than their counterparts from non-agriculture regions, what was hypothesized to be related to the increased food availability associated with agriculture (e.g., fruits, mice, and cattle breeding) (Yom-Tov et al. 2007).

Even in big cities, such as London (UK), we may find examples of species adaptability to highly artificial environments. In this European city, the mosquito *Culex pipiens* has two genetically distinct populations, adapted to different habitats: a surface population and another that inhabits the London's underground railway system. This last population is morphologically indistinguishable from the above-ground one, but adapted to the stable environment where it lives and due to the temperature stability has suppressed its diapause (Byrne and Nichols 1999). Other polluted regions, as mines or areas affected by industrial wastes, have molded resident species. Populations have developed genetic-based strategies to overcome these environments (e.g., *Limnodrilus hoffmeisteri* adaptation to metal-polluted sites in New York, USA—Klerks and Levinton 1989) that involve in the development of detoxification mechanisms, such as the deposition of cadmium in intracellular granules or the formation of organic complexes (e.g., *Porcellio scaber* from zinc–lead mine in Belgian and zinc smelter in the Netherlands—Donker and Bogert 1991) or greater growth rate, which results in larger body size that, due to the smaller surface-to-volume ratio and developmental period inside contaminated water, enhances the physiological tolerance (e.g., *Rana arvalis* in Sweden—Räsänen et al. 2003).

Overexploited populations by commercial fishing have genetically shifted their reproduction strategies to handle the extra predatory pressure. An experiment that mimics the predatory impact of commercial fishing, using guppies (*Poecilia reticulata*) as the model species, showed that animals subject to high predation presented earlier maturity, higher number, and smaller size of offspring, resulting in a rapid production of young that may overcome high predation. These traits are maintained along several generations indicating a genetic basis (Reznick and Ghalambor 2005).

Finally, some examples also involve species' ecological adaptations. House finches (*Carpodacus mexicanus*) from California were introduced in Long Island (New York, USA) in the 1950s. In their native range, the species is usually considered sedentary. However, in the new eastern range, the species has developed a latitudinal migration due to seasonal climate variations, not found in California (Able and Belthoff 1998). The development of this migratory behavior results from the interaction between the physical and social environmental characteristics with the individual's genetic structure.

Other species have genetically changed their food preferences. For example, the butterfly, *Euphydryas editha*, has reacted to the introduction of the European weed (*Plantago lanceolata*) by farmers for cattle, by incorporating these plants in its diet, and in some cases even preferring this new food resource, demonstrating a genetic change in preferences (Singer et al. 1993).

The introduction of an alien predator may also genetically shape the native preys' anti-predatory behavior. In Oregon (USA), the introduction of *Rana catesbeiana* has led the populations of *Rana aurora* tadpoles to develop genetic basis anti-predator strategies when exposed to the predators' chemical signs, such as movement reduction and increased use of shelter (Kiesecker and Blaustein 1997).

4.4 Conclusions

Although we have presented several examples of acclimation and adaptation processes of species trying to survive in a changing environment, there is a threshold of changes that a species may manage to overcome. This limit is species specific and directly related to the species natural history, plasticity and genetic structure and diversity. Acclimation and adaptation processes make it possible for species to survive in changing environments, and therefore, their knowledge should be the base of any conservation plan.

Conservation biology has long been focused on the species distribution, biodiversity, and abundance patterns (e.g., Mittermeier et al. 1998) and, more recently, on genetic evolutionarily significant unit (ESU) (Moritz 1994). This later approach has led conservation efforts to characterize intra- and interpopulation genetic variation, to develop methodologies that allow preserving genetic variability and to minimize the effects of variability loss (Stockwell et al. 2003). However, maintaining biodiversity is only possible if to the preservation of the biodiversity patterns, we add the protection of those processes that maintain, sustain, and generate biodiversity (Rouget et al. 2003). Particularly in man-shaped environments, the increasing anthropic selection pressure on wildlife implies that the focus of conservation should rapidly be concentrated on protecting ecological and evolutionary processes and identify their spatial dimensions (Rouget et al. 2003). For example, in South Africa, Rouget et al. (2003) suggested that the conservation of the Cape Floristic Region (a biodiversity hot spot, where 30 % has been converted to agriculture or urban environments) should be more effective if preservation efforts are focused on areas with high ecological adaptive diversification (e.g., ecotones; preserving ecological processes), together with others containing historically isolated populations (thus preserving evolutionary processes). Thus, it is clear that efficient conservation approaches should equally consider ecological processes together with evolutionary processes involving genetic variation in adaptive significance (Crandall et al. 2000), even in relatively short period of time (Levin 1968).

References

Able KP, Belthoff JR (1998) Rapid 'evolution' of migratory behaviour in the introduced house finch of eastern North America. Proc Roy Soc Lond B Biol Sci 265:2063–2071. doi:10.1098/rspb.1998.0541

Barnett AA, Read N, Scurlock J, Low C, Norris H, Shapley R (2000) Ecology of rodent communities in agricultural habitats in eastern Sierra Leone: Cocoa groves as forest refugia. Trop Ecol 41:127–142

Blondel J (2006) The 'design' of Mediterranean landscapes: a millennial story of Humans and ecological systems during the historic period. Hum Ecol 34:713–729. doi:10.1007/s10745-006-9030-4

Bodmer RE, Eisenberg JF, Redford KH (1997) Hunting and the likelihood of extinction of Amazonian Mammals. Conserv Biol 11:460–466

Briones V, Juan L, Sánchez C, Vela AI, Galka M, Montero N, Goyache J, Aranaz A, Mateos A, Domínguez L (2000) Bovine tuberculosis and the endangered lynx. Emerg Infect Dis 6:189–191

Byrne K, Nichols RA (1999) *Culex pipiens* in London underground tunnels: differentiation between surface and subterranean populations. Heredity 82:7–15

Campos CB, Esteves CF, Ferraz KMPMB, Crawshaw PG Jr, Verdade LM (2007) Diet of free-ranging cats (*Felis catus*) and dogs (*Canis familiares*) in a suburban and rural environment of Southeastern Brazil. J Zool 273(1):14–20. doi:10.1111/j.1469-7998.2007.00291.x

Caro T (2007) The pleistocene re-wilding gambit. Trends Ecol Evol 22:281–283. doi:10.1016/j.tree.2007.03.001

Carroll SP, Boyd C (1992) Host race radiation in the soapberry bug natural history, with the history. Evolution 46:1052–1069

Carroll SP, Klassen SP, Dingle H (1998) Rapidly evolving adaptations to host ecology and nutrition in the soapberry bug. Evol Ecol 12:955–968

Chevin LM, Lande R, Mace GM (2010) Adaptation, plasticity, and extinction in a changing environment: towards a predictive theory. PLoS Biol 8:e1000357. doi:10.1371/journal.pbio.1000357

Clutton-Brock J (1992) The process of domestication. Mammal Rev 22:79–85

Collares-Pereira MJ, Cowx I, Ribeiro F, Rodrigues J, Rogado L (2000) Threats imposed by water resource development schemes on the conservation of endangered fish species in the Guadiana River basin in Portugal. Fish Manage Ecol 7:167–178

Colles A, Liow LH, Prinzing A (2009) Are specialists at risk under environmental change? Neoecological, paleoecological and phylogenetic approaches. Ecol Lett 12:849–863. doi:10.1111/j.1461-0248.2009.01336.x

Colwell RK, Dunn RR, Harris NC (2012) Coextinction and persistence of dependent species in a changing world. Annu Rev Ecol Evol Syst 43:183–203. doi:10.1146/annurev-ecolsys-110411-160304

Crandall KA, Bininda-Emonds ORP, Mace GM, Wayne RK (2000) Considering evolutionary processes in conservation biology. Trends Ecol Evol 15:290–295. doi:10.1016/S0169-5347(00)01876-0

Descola P (1996) Constructing natures: symbolic ecology and social practice. In: Descola P, Pálsson G (eds) Nature and society: anthropological perspectives. Routledge, New York, pp 82–102

Donker MH, Bogert CG (1991) Adaptation to cadmium in three populations of the isopod *Porcellio scaber*. Comp Biochem Physiol C Comp Pharmacol 100:143–146

Driscoll DA (2004) Extinction and outbreaks accompany fragmentation of a reptile community. Ecol Appl 14:220–240. doi:10.1890/02-5248

Eterovic A, Duarte MR (2002) Exotic snakes in São Paulo City, southeastern Brazil: why xenophobia? Biodivers Conserv 11:327–339. doi:10.1023/a:1014509923673

FAO—Food and Agriculture Organization (2012) http://www.fao.org/docrep/005/X4988E/x4988e01.htm. Cited 17 Nov 2012

Feare CJ, Wadsworth JT (1981) Starling damage on farms using the complete diet system of feeding dairy cows. Anim Prod 32:179–183

Fedriani JM, Fuller TK, Sauvajot RM (2001) Does availability of anthropogenic food enhance densities of omnivorous mammals? An example with coyotes in southern California. Ecography 24:325–331. doi:10.1034/j.1600-0587.2001.240310.x
Ferraz KMPMB, Ferraz SFB, Moreira JR, Verdade LM (2007) Capybara (*Hydrochoerus hydrochaeris*) distribution in agroecosystems: a cross-scale habitat analysis. J Biogeogr 33:223–230. doi:10.1111/j.1365-2699.2006.01568.x
Ferrer M, Negro JJ (2004) The near extinction of two large European predators: super specialists pay a price. Conserv Biol 18:344–349. doi:10.1111/j.1523-1739.2004.00096.x
Fisher DO, Blomberg SP, Owens IPF (2003) Extrinsic versus intrinsic factors in the decline and extinction of Australian marsupials. Proc Roy Soc Lond B Biol Sci 270:1801–1808. doi:10.1098/rspb.2003.2447
Gheler-Costa C, Verdade LM, Almeida AF (2002) Mamíferos não-voadores do campus "Luiz de Queiroz", Universidade de São Paulo, Piracicaba, Brasil. Rev Bras Zool 19:203–214
Gheler-Costa C, Vettorazzi CA, Pardini R, Verdade LM (2012) The distribution and abundance of small mammals in agroecosystems of Southeastern Brazil. Mammalia 76:185–191. doi:10.1515/mammalia-2011-0109
Goszczynski J, Jedrzejewska B, Jedrzejewski W (2000) Diet composition of badgers (*Meles meles*) in a pristine forest and rural habitats of Poland compared to other European populations. J Zool 250:495–505
Gravel M, Mazerolle MJ, Villard MA (2012) Interactive effects of roads and weather on juvenile amphibian movements. Amphibia-Reptilia 33:113–127. doi:10.1163/156853812x625512
Greville RW, Morgan AJ (1991) A comparison of Pb, Cd and Zn accumulation in terrestrial slugs maintained in microcosms: Evidence for metal tolerance. Environ Pollut 74:115–127
Groman JD, Pellmyr O (2000) Rapid evolution and specialization following host colonization in a yucca moth. J Evol Biol 13:223–236. doi:10.1046/j.1420-9101.2000.00159.x
Hendry AP, Quinn TP (1997) Variation in adult life history and morphology among Lake Washington sockeye salmon (*Oncorhynchus nerka*) populations in relation to habitat features and ancestral affinities. Can J Fish Aquat Sci 54:75–84
Hendry AP, Wenburg JK, Bentzen P, Volk EC, Quinn TP (2000) Rapid evolution of reproductive isolation in the wild: evidence from introduced salmon. Science 290:516–518
Holdway DA (2002) The acute and chronic effects of wastes associated with offshore oil and gas production on temperate and tropical marine ecological processes. Mar Pollut Bull 44:185–203. doi:10.1016/S0025-326X(01)00197-7
Kiesecker JM, Blaustein AR (1997) Population differences in responses of red-legged frogs (*Rana aurora*) to introduced bullfrogs. Ecology 78:1752–1760
King CM, Moors PJ (1979) The life-history tactics of mustelids, and their significance for predator control and conservation in New Zealand. New Zeal J Zool 6:619–622
Klerks PL, Levinton JS (1989) Rapid evolution of metal resistance in a benthic oligochaete inhabiting a metal-polluted site. Biol Bull 176:135–141
Kumar MA, Singh M (2011) Behavior of Asian elephant (*Elephas maximus*) in a land-use mosaic: conservation implications for human-elephant coexistence in the Anamalai hills, India. Wildl Biol Pract 6:69–80
Kupfer JA, Franklin SB (2009) Linking spatial pattern and ecological responses in Human-modified landscapes: the effects of deforestation and forest fragmentation on biodiversity. Geogr Compass 3:1331–1355. doi:10.1111/j.1749-8198.2009.00245.x
Levin R (1968) Evolution in changing environments. Princeton University Press, Princeton
Lindenmayer D, Burgman M (2005) Practical conservation biology. CSIRO Publishing, Collingwood
Lomolino MV, Perault DR (2007) Body size variation of mammals in a fragmented, temperate rainforest. Conserv Biol 21:1059–1069. doi:10.1111/j.1523-1739.2007.00727.x
Luniak M (2004) Synurbization—adaptation of animal wildlife to urban development. In: Shaw WW, Harris L (eds) Proceedings of the fourth international symposium on urban wildlife conservation. University of Tucson, Tucson, pp 50–55

Mann RM, Hyne RV, Choung CB, Wilson SP (2009) Amphibians and agricultural chemicals: review of the risks in a complex environment. Environ Pollut 157:2903–2927. doi:10.1016/j.envpol.2009.05.015

Marques C, Rosalino LM, Santos-Reis M (2007) Otter predation in a trout fish farm of central–East Portugal: preference for 'fast-food'? River Res Appl 23:1147–1153. doi:10.1002/rra.1037

Martin PS, Gheler-Costa C, Lopes PC, Rosalino LM, Verdade LM (2012) Terrestrial non-volant small mammals in agro-silvicultural landscapes of Southeastern Brazil. For Ecol Manage 282:185–195. doi:10.1016/j.foreco.2012.07.002

Mittermeier RA, Myers N, Thomsen JB, Da Fonseca GAB, Olivieri S (1998) Biodiversity hotspots and major tropical wilderness areas: Approaches to setting conservation priorities. Conserv Biol 12:516–520

Moritz C (1994) Defining 'evolutionary significant units' for conservation. Trends Ecol Evol 9:373–375

Naveh Z (1975) The evolutionary significance of fire in the Mediterranean region. Vegetatio 29:199–208

Negro JJ, Hiraldo F (1993) Nest-site selection and breeding success in the Lesser Kestrel *Falco naumanni*. Bird Study 40:115–119

Owens IPF, Bennet PM (2000) Ecological basis of extinction risk in birds: habitat loss versus human persecution and introduced predators. Proc Natl Acad Sci USA 97:12144–12148

Palmeira FBL, Crawshaw PG Jr, Haddad CM, Ferraz KMPMB, Verdade LM (2008) Cattle depredation by puma (*Puma concolor*) and jaguar (*Panthera onca*) in Northern Goiás, Central-western Brazil. Biol Conserv 141:118–125. doi:10.1016/j.biocon.2007.09.015

Palomo LJ, Gisbert J (2002) Atlas de los mamíferos terrestres de España. Dirección General de Conservación de la Naturaleza-SECEM-SECEMU, Madrid

Pertoldi C, García-Perea R, Godoy JA, Delibes M, Loeschcke V (2006) Morphological consequences of range fragmentation and population decline on the endangered Iberian lynx (*Lynx pardinus*). J Zool 268:73–86. doi:10.1111/j.1469-7998.2005.00024.x

Phillips BL, Shine R (2004) Adapting to an invasive species: toxic cane toads induce morphological change in Australian snakes. Proc Natl Acad Sci USA 101:17150–17155

Pinto-Correia T, Vos W (2004) Multifunctionality in Mediterranean landscapes—past and future. In: Jongman R (ed) The new dimensions of the European landscape Springer, Dordrecht, pp 135–164

Piña CI, Lance VA, Ferronato BO, Guardia I, Marques TS, Verdade LM (2009) Heavy metal contamination in *Phrynops geoffroanus* (Schweigger, 1812) (Testudines: Chelidae) in a river basin, São Paulo, Brazil. Bull Environ Contam Toxicol 83:771–775

Potts R (2004) Environmental variability and its impact on adaptive evolution, with special reference to human origins. In: Rothschild L, Lister A (eds) Evolution on Planet Earth. Academic Press, London, pp 363–378

Preston PW (1960) Time and space and the variation of species. Ecology 41:612–627

Räsänen K, Laurila A, Merilä J (2003) Geographic variation in acid stress tolerance of the moor frog. *Rana arvalis*. I. Local adaptation. Evolution 57:352–362. doi:10.1111/j.0014-3820.2003.tb00269.x

Real R, Barbosa AM, Rodríguez A, García FJ, Vargas JM, Palomo LJ, Delibes M (2009) Conservation biogeography of ecologically interacting species: the case of the Iberian lynx and the European rabbit. Divers Distrib 15:390–400. doi:10.1111/j.1472-4642.2008.00546.x

Reznick DN, Ghalambor CK (2005) Can commercial fishing cause evolution? Answers from guppies (*Poecilia reticulata*). Can J Fish Aquat Sci 62:791–801. doi:10.1139/f05-079

Rosalino LM, Ferreira D, Leitão I, Santos-Reis M (2011) Usage patterns of Mediterranean agro-forest habitat components by wood mice *Apodemus sylvaticus*. Mamm Biol 76:268–273. doi:10.1016/j.mambio.2010.08.004

Rosalino LM, Loureiro F, Macdonald DW, Santos-Reis M (2005) Dietary shifts of the badger (*Meles meles*) in Mediterranean woodlands: an opportunistic forager with seasonal specialisms. Mamm Biol 70:12–23. doi:10.1078/1616-5047-00172

Rosalino LM, Martin PS, Gheler-Costa C, Lopes PC, Verdade LM (2013) Allometric relations of Neotropical small rodents (Sigmodontinae) in anthropogenic environments. Zool Sci 30: 585–590. doi:10.2108/zsj.30.585

Rouget M, Cowling RM, Pressey RL, Richardson DM (2003) Identifying spatial components of ecological and evolutionary processes for regional conservation planning in the Cape Floristic Region, South Africa. Divers Distrib 9:191–210. doi:10.1046/j.1472-4642.2003.00025.x

Safi K, Kerth G (2004) A comparative analysis of specialization and extinction risk in temperate-zone bats. Conserv Biol 18:1293–1299. doi:10.1111/j.1523-1739.2004.00155.x

Schmidt NM, Ims RA, Høye TT, Gilg O, Hansen LH, Hansen J, Lund M, Fuglei E, Forchhammer MC, Sittler B (2012) Response of an arctic predator guild to collapsing lemming cycles. Proc Biol Sci 279:4417–4422. doi:10.1098/rspb.2012.1490

Silva JS (2007) Árvores e florestas de Portugal: Pinhais e eucaliptais, a floresta cultivada. Edições Público, Lisboa

Singer MC, Thomas CD, Parmesan C (1993) Rapid human-induced evolution of insect-host associations. Nature 366:681–683

Stockwell CA, Hendry AP, Kinnison MT (2003) Contemporary evolution meets conservation biology. Trends Ecol Evol 18:94–101

Thompson JN (1998) Rapid evolution as an ecological process. Trends Ecol Evol 13:329–332

Tóth M, Bárány A, Kis R (2009) An evaluation of stone marten (*Martes foina*) records in the city of Budapest, Hungary. Acta Zool Acad Sci Hung 55:199–209

Travis JMJ (2003) Climate change and habitat destruction: a deadly anthropogenic cocktail. Proc Roy Soc Lond B Biol Sci 270:467–473. doi:10.1098/rspb.2002.2246

Verdade LM (1996) The influence of hunting pressure on the social behavior of vertebrates. Rev Bras Biol 56:1–13

Verdade LM, Campos CB (2004) How much is a puma worth? Economic compensation as an alternative for the conflict between wildlife conservation and livestock production in Brazil. Biota Neotropica 4(2):1–4

Verdade LM, Ferraz KMPMB (2006) Capybaras (*Hydrochoerus hydrochaeris*) on an anthropogenic habitat in Southeastern Brazil. Braz J Biol 66:371–378

Verdade LM, Rosalino LM, Gheler-Costa C, Pedroso NM, Lyra-Jorge MC (2011) Adaptation of mesocarnivores (Mammalia: Carnivora) to agricultural landscapes of Mediterranean Europe and Southeastern Brazil: a trophic perspective. In: Rosalino LM, Gheler-Costa C (eds) Middle-sized carnivores in agricultural landscapes. Nova Science Publishers, New York, pp 1–38

Vos J (2000) Food habits and livestock depredation in two Iberian wolf packs (*Canis lupus signatus*) in the north of Portugal. J Zool 251:457–462

Wayland M, Crosley R (2006) Selenium and other trace elements in aquatic insects in coal mine–affected streams in the rocky mountains of Alberta, Canada. Arch Environ Contam Toxicol 50:511–522. doi:10.1007/s00244-005-0114-8

Wegner J, Merriam G (1990) Use of spatial elements in a farmland mosaic by a woodland rodent. Biol Conserv 54:263–276

Williams CK, Moore RJ (1989) Phenotypic adaptation and natural selection in the wild rabbit, *Oryctolagus cuniculus*, in Australia. J Anim Ecol 58:495–507

Yom-Tov Y, Yom-Tov S, Barreiro J, Blanco JC (2007) Body size of the red fox *Vulpes vulpes* in Spain: the effect of agriculture. Biol J Linn Soc 90:729–734

Yamada F, Sugimura K (2004) Negative impact of an invasive small Indian mongoose *Herpestes javanicus* on native wildlife species and evaluation of a control project in Amami-Ohshima and Okinawa Islands, Japan. Global Environ Res 8:117–124

Chapter 5
Biodiversity Loss and Infectious Diseases

Kevin D. Lafferty

Abstract When conservation biologists think about infectious diseases, their thoughts are mostly negative. Infectious diseases have been associated with the extinction and endangerment of some species, though this is rare, and other factors like habitat loss and poorly regulated harvest still are the overwhelming drivers of endangerment. Parasites are pervasive and play important roles as natural enemies on par with top predators, from regulating population abundances to maintaining species diversity. Sometimes, parasites themselves can be endangered. However, it seems unlikely that humans will miss extinct parasites. Parasites are often sensitive to habitat loss and degradation, making them positive indicators of ecosystem "health". Conservation biologists need to carefully consider infectious diseases when planning conservation actions. This can include minimizing the movement of domestic and invasive species, vaccination, and culling.

5.1 Introduction

We have all been sick from infectious diseases, and this predisposes us to view parasites with disdain. Here, I discuss the importance of infectious diseases (i.e., parasites and pathogens) for conservation. This is not a common topic. Nearly, half of conservation biology texts do not even mention infectious diseases (Nichols and Gómez 2011). Half of those texts that do mention infectious diseases only consider negative impacts of disease. But the story is much richer than this. Infectious diseases play important roles in ecosystems, hurting some species and favoring others. Under rare circumstances, they can cause their hosts to become

K. D. Lafferty (✉)
U.S. Geological Survey, Western Ecological Research Center, Marine Science Institute, University of California, Santa Barbara, CA 93106, USA
e-mail: klafferty@usgs.gov

L. M. Verdade et al. (eds.), *Applied Ecology and Human Dimensions in Biological Conservation*, DOI: 10.1007/978-3-642-54751-5_5,

endangered. Environmental change can favor or impair infectious diseases. Sometimes, parasites themselves can be endangered. For these reasons, parasite can give us interesting insight into environmental degradation, making them useful indicators. Furthermore, there are several actions that conservation biologists can take to protect species at risk from infectious disease.

5.2 The Role of Disease in Ecosystems

Parasites are pervasive. But due to their small size, parasites seem insignificant players at the ecosystem level. Do their numbers add up? Can they have effects even greater than their numbers would imply? What are their contributions to biodiversity and food webs? When do they control host populations?

Parasitism is a popular lifestyle, but exactly how popular is hard to tell because parasitologists have not yet looked at most animal species. What information exists about parasites is often only from one location and rarely for all parasite groups. Some authors have estimated the proportion of described species in various animal taxa that are parasitic. Poulin and Morand (2004) estimated that there were about 1.5 parasite species per vertebrate species. Several molecular genetic studies suggest that described parasite species are often suites of cryptic species that are simply difficult to distinguish morphologically (Miura et al 2005). If cryptic species are more common for parasites than for free-living species (highly possible given the lack of morphological characters in some parasite groups), there could be an even higher proportion of parasites on earth. An alternative approach is to go to a particular system and to estimate the richness of free-living and parasitic species. This has been done for estuarine systems where a third of the 314 species encountered are parasites, and this is probably a gross underestimate (Hechinger et al 2011b). It is unknown whether this percentage of parasitism is representative of other types of ecosystems, but parasites are unquestionably a large part of biodiversity.

Parasites are embedded in food webs, which track the flow of energy through ecosystems and are a fundamental theme of ecology. Even though many parasites are host specific, overall, parasites tend to have more hosts than predators have prey (Lafferty et al 2006). In part, this is due to complex life cycles, for which parasites can have one or more hosts per stage (Rudolf and Lafferty 2011). It is less commonly realized that consumers eat parasites, either when the parasites are larvae, or incidentally when parasites are inside prey (Johnson et al 2010). Inclusion of parasites in food webs greatly alters food-web structure, increasing measures like connectance and nestedness (Lafferty et al 2006). In addition, parasites make food webs less robust, because, as will be discussed below, parasites are more likely to suffer secondary extinctions than are free-living species (Lafferty and Kuris 2009; Rudolf and Lafferty 2011). So, from a food-web perspective, parasites appear to be important players in ecosystems (Lafferty et al 2008a).

One way to consider the role of a species in an ecosystem is to measure its biomass density. Although parasites are integral parts of food webs, for them to affect the flow of energy through a system, they must make up some biomass of that system. Parasites are small, so they might not be as important energetically as larger, free-living species. Kuris et al (2008) found that parasites make up 1–2 % of the living biomass in three estuaries. Although 1–2 % might not seem like much, it is exactly what is expected once one accounts for the upper trophic levels at which parasites operate (Hechinger et al 2011a). For instance, in three well-sampled estuaries, there was a greater biomass density of trematode parasites in snails than there was biomass density of birds (Kuris et al 2008). This means there is no reason to suspect that parasites have any less of a role in ecosystems than top predators.

Parasites might reduce or regulate the abundance of their hosts (Tompkins and Begon 1999). To reduce host abundance, exposure to parasites needs to be common, and parasites need to negatively affect infected hosts. For instance, adding tapeworm eggs to beetle colonies depresses the density of the beetle host because infected beetles have decreased fecundity and survivorship (Keymer 1982). To regulate the host population, however, means to reduce host density when hosts are abundant, but not when hosts are rare. Host population regulation is a key assumption that affects how we hypothesize the role infectious diseases in conservation biology. Modeling has helped determine the criteria under which pathogens and parasites can regulate host populations (Anderson and May 1978; May and Anderson 1978). The basic premise behind regulation for pathogens is density-dependent transmission. Such pathogens only invade dense host populations and fade out when the density of susceptible hosts declines (Lloyd-Smith et al 2005). However, for typical parasites, hosts can be re-infected, but density-dependent effects (e.g., crowding or increased mortality for heavily infected hosts) will limit parasite abundance. Additional complexities such as refuges from parasitism or invulnerable stages can theoretically allow the host to persist when the infectious agent becomes common. In a classic experiment on parasite regulation of host abundance, a nematode parasite was able to depress the population of laboratory mice to 10 % of control densities (Scott 1987). Examples of regulation (or failed regulation) from the field include the whole of biological control literature. The dramatic effect of myxomatosis virus on rabbits in Australia is a textbook example (Fenner and Ratcliffe 1965). Whereas myxomatosis caused mass mortalities of rabbits, a nematode parasite that reduces the fecundity of reindeer appears to reduce host densities to a relatively stable equilibrium (Albon et al 2002). A potential consequence of regulation by infectious diseases is the promotion of biodiversity. If infectious diseases prevent some species from out-competing others, coexistence among competitors becomes more likely (Clay et al 2008). Whether or not parasites "control" host populations, most host populations would be more abundant if they did not suffer from infectious diseases.

Some parasites manipulate their hosts, with consequences for conservation. Parasites can increase the susceptibility of their intermediate hosts to predation by final hosts (Lafferty 1999), and this can alter predator prey dynamics (Dobson 1988; Lafferty 1992). For instance, mathematical models suggest that a tapeworm that debilitates moose might allow endangered wolves to persist in some locations (Hadeler and Freedman 1989). A recent example indicates how a manipulative parasite can have a positive indirect benefit for conservation. The manipulating parasite is a nematomorph worm that causes its cricket host to jump into streams where the worm reproduces (Thomas et al 2002). In Japan, these manipulated crickets form the bulk of the diet for an endangered trout (Sato et al 2011). Without the parasite, these trout might become extinct. In these and other examples, the parasite benefits predators and impacts prey populations. It would be useful to know whether other endangered predators receive indirect benefits from parasites and how this might be used in management programs.

Parasites make up much of biodiversity, and they appear to play important roles. They are common parts of food webs with many connections to free-living species. Though small, when combined, they have as much mass as predator populations. Parasites have the potential to affect species of concern, because they can depress host populations. However, due to density-dependent transmission, the effect of parasites will tend to wane as hosts become rare. This can lead parasites to handicap competitive dominants, facilitating biodiversity and coexistence. Although the direct effects of parasites are bad for host individuals, indirect effects might be positive, particularly for predators that feed on prey manipulated by parasites. Biologists have been wrong to ignore the role of parasites in natural systems, but this is changing for the better.

5.3 Diseases as Agents of Endangerment

Some parasites can kill or seriously affect the health of their hosts, and a few can have noticeable effects on host populations. Anthrax, plague, influenza, HIV, small pox, malaria, hookworm, river blindness, and dysentery are examples of infectious diseases that have shaped human history. When might an infectious disease endanger its host, or cause its extinction? Which types of infectious diseases are more commonly associated with conservation impacts?

Exceptions to basic epidemiological theory must occur for an infectious disease to extirpate a host (de Castro and Bolker 2005; Lafferty and Gerber 2002). Under typical density-dependent transmission, as disease drives host populations down, it crosses a threshold density, below which the parasite can no longer transmit fast enough to persist in the host population. Nevertheless, some circumstances can prevent disease fade out. For instance, captive breeding programs maintain animals at high densities in association with other species. In a notable example, a captive colony of black-footed ferrets was nearly extirpated when the group was accidentally exposed to canine distemper virus (CDV) (Williams et al 1988). Similarly, the last known *Partula turgida* land snails from Tahiti (Cunningham and Daszak 1998) were extirpated from the London Zoo after a microsporidian pathogen contaminated the cultures. Although this is the first documented extinction caused by a parasite, the snail's earlier extinction in the wild was caused by the introduction of a predatory snail. In nature, an infectious disease can extirpate a host if it has a second, more tolerant, host species. American gray squirrels have replaced British red squirrels, in part due to a shared parapox virus introduced with the tolerant gray squirrel (Tompkins et al 2002). Likewise, canine distemper from domestic dogs can spillover to endangered wolves, lynxes, wild dogs, foxes, and lions, causing heavy mortality (Cleaveland 2009). Alternatively, if the disease agent can live outside the host, it will be able to survive periods of low host abundance, and not fade out. For instance, the chytrid fungus that causes mass mortalities in some species of amphibians can grow saprophytically without amphibians (Longcore et al 1999). Other examples are more complicated. For instance, a long time lag between infection and pathology can allow an infectious disease to reach a high prevalence before driving host numbers down (Lloyd-Smith et al 2005). In California, endangered intertidal black abalone are susceptible to a bacterial pathogen, but do not normally exhibit mortality, allowing all abalone to become infected in a local population; however, when water temperatures become warm, infected animals die, leading to mass mortalities (Ben-Horin et al 2013). Knowing the special circumstances under which infectious diseases can drive hosts to low abundances is essential when trying to manage endangered species.

Although infectious disease is listed as one of the five main causes of extinction (Wilcove et al 1998), links between disease and endangerment are not common (Smith et al 2006). In a summary of the IUCN Red List of Threatened and endangered species, Smith et al (2006) found that infectious disease was a contributing factor in <4 % of the 833 plants and animals documented to have gone extinct in modern times and <8 % of the 2,852 critically endangered species. These numbers relegate infectious disease to a relatively minor threat to species in contrast with habitat destruction and hunting. It seems likely, therefore, that the special cases that cause infectious diseases to seriously affect populations of their host species are not pervasive in nature. Still, they are numerous enough to take seriously.

Some types of infectious diseases repeatedly affect host species of concern. For initially common host species, the most common problems are introduced fungal, viral, and protozoal pathogens (e.g., chytrid fungus, avian malaria); for already endangered species, viruses that spillover from domestic animals (e.g., rabies, CDV), most commonly dogs, are the greatest concern (Lafferty and Gerber 2002; Smith et al 2006). Fungal diseases are particularly vexing because they are not well understood even though they have been an issue for conservation biologists for a long time. Chestnut blight was one of the first infectious diseases of conservation concern, whereas white-nose syndrome is a recent fungus (*Geomyces destructans*) to North America thought to be driving bat species toward extinction in the USA but not in Europe. Despite the preponderance of viruses and fungi, several other parasitic groups are of concern to conservation biologists.

Parasites are not common sources of endangerment and are even more rarely associated with extinctions. This is due to the importance of density-dependent transmission, which causes many infectious diseases to fade out before they drive their host populations to low densities. However, when an introduced or domestic species supports a pathogenic infectious disease, a spillover into wildlife or captive populations can lead to serious threats.

5.4 The Response of Infectious Diseases to Environmental Degradation

The world is changing. Human actions that endanger species and degrade the environment can also affect infectious diseases. Conservation biologists are most concerned with situations where environmental impacts also increase infectious diseases. However, infectious diseases might also suffer from impacts. The outcome at the population level should depend on how stressors interact with the vital rates of hosts and infectious diseases. Pollution, biodiversity loss, hunting/fishing, and climate change might favor some infectious diseases but impair others. Have there been changes to infectious diseases over time associated with environmental degradation?

Stress can have different effects on host populations than on individuals (Lafferty and Holt 2003). Although a stressed individual is more likely to become infected with a parasite, it is also more likely to die, thereby reducing the abundance of infected hosts and increasing the mortality rate of parasites within hosts. For this reason, the net effect of stress on a parasite population is difficult to predict and can, counter-intuitively, lead to decreases in parasitism.

Pollution can increase host susceptibility to infection, but it can also be toxic for parasites (Lafferty 1997). Many free-living parasite stages (e.g., eggs, larvae) have shorter life spans when exposed to toxic substances. Furthermore, parasites can be more susceptible to contaminants than their hosts are (this is the premise behind using drugs to treat infectious diseases). As a result, helminths tend to decline with hydrocarbon or heavy metal exposure, whereas some protozoans and monogeneans increase in polluted areas. Most parasites increase in prevalence with eutrophication, because nutrients increase the productivity of host populations. The response of parasites to pollution, therefore, depends on the type of pollution and the type of parasite.

The addition of "non-competent" species to a habitat can theoretically reduce the transmission of some types of infectious diseases via the dilution effect (Keesing et al 2006). The dilution effect is mostly likely to occur for vector-transmitted diseases, in which transmission is frequency dependent. This means that bite rates of vectors are not usually limited by host availability. In such cases, vectors can bite several different types of hosts. If a vector carrying a host-specific pathogen bites a non-competent host, the disease will not transmit. Some assumptions are needed for the dilution effect to occur: Non-competent hosts are lost from communities before competent hosts, and non-competent hosts do not magnify vector populations. The dilution effect has been touted as a win-win situation for biodiversity and human health if the addition of non-competent species is associated with increases in biodiversity (Keesing et al 2010). Despite its popularity among conservation biologists, it is not clear how often the dilution effect occurs in nature and whether dilution is positively associated with biodiversity (Salkeld et al 2013).

Although the dilution effect is a theoretical possibility, biodiversity loss can lead to the decline of some types of infectious diseases. This is particularly true when parasites are host specific and have complex life cycles. For generalist parasites, the loss of a single host species will not eliminate the parasite from the system. Parasites with complex life cycles, however, require at least one species from each obligate host category (Lafferty and Kuris 2009; Rudolf and Lafferty 2011). Such parasites can be sensitive to biodiversity loss. If biodiversity loss leads to a few abundant species, one might expect to find a few prevalent parasite species. In general, parasite diversity and abundance should follow host diversity and abundance (Hechinger and Lafferty 2005; Lafferty 2012).

Hunting and fishing are types of biodiversity loss that can reduce host abundance and thus alter disease dynamics (Dobson and May 1987; Wood et al 2010). As fishing drives target species below a threshold level for transmission, parasite species will not be able to complete their life cycles. In addition, sport fishing and certain commercial gear targets the larger, older individuals that also have the most parasites. Reports of parasites of marine mammals have increased since these animals were released from hunting pressure, whereas reports of parasites of fishes have decreased as many fish stocks crashed (Ward and Lafferty 2004). Experimental fishing drives parasites to low levels (Amundsen and Kristoffersen 1990), confirming a causal link between fishing and parasite loss. This can have community-level implications. For instance, parasite communities are more diverse in coral reef fishes at unfished sites than at fished sites (Lafferty et al 2008b). Sometimes, fishing can have indirect, positive effects on parasites. When fishing top predators releases prey populations from predation pressure, parasites of prey will benefit (Behrens and Lafferty 2004; Lafferty 2004; Packer et al 2003; Sonnenholzner et al 2011). For these reasons, it can be difficult to predict the net effect of fishing on the diseases of an ecosystem.

Climate change has the potential to alter the distribution of infectious diseases. Although disease expansion in higher latitudes gets the most attention, areas near the equator might become too warm for parasites (Lafferty 2009). As a result, some locations will see more infectious diseases, while other locations will see decreases. The biggest changes are likely to occur at high latitudes where climate is changing most rapidly and where tropical diseases can expand (Kutz et al 2005). Extreme weather events can affect hosts and parasites. For instance, a hurricane that devastated the Yucatan Peninsula, Mexico, in 2007 impacted free-living species, but it was the parasites that took the longest to recover (Aguirre-Macedo et al 2011). Overall, climate change should create similar challenges for hosts and parasite alike.

As humans degrade the environment, biodiversity will decline, both for parasites and free-living species. Those few cases where free-living species will decline, but parasites will increase, will create a special challenge to conservation biology. In particular, if climate change introduces new pathogens to naive hosts, impacts could occur. Nevertheless, overall, pollution, fishing, and climate change seem as likely to harm parasites as to benefit parasites.

5.5 Endangered Parasites

Parasites are sensitive to environmental change, and some have suggested that they could make up the unseen majority of species extinctions (Dobson et al 2008; Dunn et al 2009; Koh et al 2004; Poulin and Morand 1997; Sprent 1992). The success of vector control in suppressing human diseases underscores how removing a host (e.g., a mosquito) can lead to parasite loss (e.g., malaria). Parasite endangerment should relate to host endangerment, host specificity, and life cycle complexity.

When all hosts are gone, no parasites can remain. For instance, the trematode *Pleurogonius malaclemys* only infects snails in the presence of the endangered diamondback terrapin (*Malaclemys terrapin*), the sole final host for the trematode (Byers et al 2011). When a diamondback terrapin population is extirpated, it takes its host-specific parasites with it. This is consistent with the observation that extinction of the snail *Cerithidea californica* is linked to the loss of several parasite species of birds that require the snail as a first intermediate host (Torchin et al 2005). The pygmy hog-sucking louse (*Haematopinus oliveri*) is specific to an endangered pig, leading it to be the only parasite listed on the IUCN Red List (Whiteman and Parker 2005). However, to my knowledge, there is no documentation of an accidental parasite extinction. An example of parasite extinction that has since been proven false is feather lice (genus *Columbicola*) from the extinct passenger pigeon. These lice were less host specific than initially thought and have been found on other species (Dunn 2002). Still, endangered species have parasites, and, if these parasites are host specific, the parasites are arguably more endangered than their hosts. Parasites can go extinct well before their hosts, because some parasites occur only in part of the range of their hosts and, for parasites with density-dependent transmission, the host only need drop below a threshold density for the parasite to go extinct. Many endangered species might have already dropped below that threshold for some of their parasites. Perhaps as a result, endangered primates have fewer parasites than primate species that are not threatened (Altizer et al 2007). Alternatively, such a pattern could occur if the factors that lead to host endangerment (insular, isolated populations) also limit parasite communities. In other words, if hosts with high extinction risk have fewer parasites to start with, then fewer parasites will be found in endangered species. For instance, parasites are less diverse in hosts with narrow diets (Chen et al 2008; Vitone et al 2004), and specialists should be more prone to extinction (Purvis et al 2000). On the other hand, large species, which are more likely to be threatened by habitat loss and overharvest (Purvis et al 2000), tend to host more parasite species (Vitone et al 2004). Similarly, top predators are more likely to go extinct, and parasite diversity increases with host trophic level (Lafferty et al 2006). Unfortunately, due to a lack of historical information on parasites before the biodiversity crisis, it is difficult to know which parasites have already gone extinct. No matter the association between parasite richness and host propensity of endangerment, it remains that the most endangered species are parasites of currently endangered species (Dunn et al 2009).

Host specificity makes parasites more susceptible to endangerment. One way to understand this effect is to take a set of host species and their parasites and then plot how the proportion of extant parasites would change as host species were removed from the system. The endpoints of this relationship are obvious. When all hosts are present, all parasite species are present and when no host species are present, no parasites will be present. If parasites are strictly host specific, the relationship should be linear. However, the more general the parasites are, the less sensitive they will be to biodiversity loss of hosts, leading to an initially slow parasite loss rate as hosts are removed (Koh et al 2004; Lafferty 2012). This is because generalist parasite species should be able to persist even if only a single host species remains. Some parasite taxa are more specific than others, suggesting that patterns of parasite extinction will vary from group to group. For instance, using data for North American carnivores, lice are much more host specific and, therefore, much more prone to extinction, than are viruses (Dunn et al 2009).

Life cycle complexity makes parasites more susceptible to endangerment because there are more weak links in the chain. In other words, if even one stage in the life cycle cannot find a host, the parasite cannot persist. When trying to estimate extinction risk it is, therefore, important to distinguish between a parasite that can use many hosts from a parasite that must use many different hosts (Lafferty and Kuris 2009; Rudolf and Lafferty 2011). Most parasite species require more than one host, and incorporating this into plots of parasite and host extinction opens up the possibility that proportional rates of parasite extinction can exceed proportional rates of host extinction (Lafferty 2012). In other words, the loss of parasite diversity could exceed rates of free-living extinctions, and this loss would be especially high for parasites with complex life cycles.

It seems probable that many parasites have gone extinct and multitudes more are endangered. Host-specific parasites, particularly those with complex life cycles, are most at risk. Although few will shed a tear at the extinction of a parasite, there are two reasons one might miss these parasites. As indicated above, parasites are important players in natural ecosystems, and losing a parasite could have just as big an ecological consequence as losing a top predator. Second, many parasites are as inherently interesting as the charismatic megafauna whose photos adorn the brochures of conservation organizations. Most people would agree we have an obligation to preserve the sperm whale (*Physeter macrocephalus*) if only because it is charismatic, large, and fierce. However, consider *Placentonema gigantissima*, a parasite of sperm whales. This species is even more endangered than the sperm whale. The worm is also fascinating in its own right, reaching over 10 m. Fortunately, saving endangered parasites takes little additional effort because conserving endangered hosts is the best way to conserve parasites. Perhaps the more relevant question is whether we should place more emphasis on protecting hosts with many host-specific parasites than hosts with none. After all, hosts with many parasites represent a trove of biodiversity.

5.6 Parasites as Indicators of Environmental Quality

Parasites can indicate impacts to the environment. A community of parasites reveals something about the community of hosts present in the environment, and knowing what parasites increase or decrease with environmental change helps make it possible to understand how the environment is changing for hosts in the system. What types of parasites make proverbial canaries in a coal mine? Ideal parasite indicators are species with complex life cycles for which one of the hosts is easy to sample. Parasites of fishes and snails show particular promise as indicators (Lafferty 1997).

Indicator species should be easy to sample and yield information about processes that are otherwise hard to come by. If you wanted to know whether a species was in decline, it might be useful to look at its parasites. This is because fewer parasites might indicate that the density of the host has dipped below the level where transmission is efficient. However, sampling endangered hosts to look for their parasites is both difficult and defeats the purpose of conserving them. Instead, one could sample an intermediate host that shares parasites with the endangered host. For this reason, parasites with complex life cycles can make good indicators of a target host if one of the other hosts in the life cycle is easy to sample (Huspeni et al 2005).

Parasites of fishes can increase or decrease with pollution, providing a biologically meaningful indicator of water quality (Marcogliese 2005). They can also indicate food-web links (Valtonen et al 2010). For instance, sharks are hard to sample and in decline. They have tapeworms with complex life cycles that use reef fishes as intermediate hosts. In areas with many sharks, the reef fishes are commonly infected with tapeworm larvae, suggesting that by sampling tapeworm larvae in small fish that are easy to collect, one could track the abundance of sharks in time and space (Lafferty et al 2008b). Many fishes are relatively easy to sample, and they have communities of parasites that reflect the food webs they live in, give information about fish density, and also indicate aspects of the physical environment.

Communities of trematode parasites in snails are another system with broad application as indicators. The final hosts of these parasites are vertebrates such as birds. Areas with a high abundance and richness of birds have a high abundance and richness of trematodes infecting local snail populations (Hechinger and Lafferty 2005). Snails from degraded portions of estuaries are less parasitized, and parasitism increases after habitat restoration, presumably because restoration makes conditions more attractive for birds that then spread trematodes to snails (Huspeni and Lafferty 2004). Trematodes and snails are broadly distributed in aquatic habitats and could be used as indicators in many locations (Huspeni et al 2005).

The use of parasites as indicators is in its infancy. It is counter-intuitive that parasites indicate good environmental conditions. Although current examples are from aquatic systems, parasites of terrestrial animals might also make good indicators. Despite the considerable evidence, it is still difficult to convince people that a healthy system is one with many parasites (Hudson et al 2006).

5.7 Conservation Strategies in a World Full of Parasites

Conservation biologists need to be aware of infectious diseases when managing threatened species (Gerber et al 2005; Lafferty and Gerber 2002). For instance, nature reserves should be designed with attention to maintaining natural infectious processes, and wildlife should be assessed for background parasites so that new diseases are easier to detect. Species invasions should be minimized to reduce the threat of novel diseases. Vaccinations and culling can be used to break the cycle of transmission. Population viability analysis (PVA) models should include infectious processes. In addition, captive rearing programs should take care to avoid exposing threatened species to disease.

Parasites are part of natural systems, and it makes sense to include them in the design of reserves, reintroductions, and conservation programs. This is not current practice. A common precaution for reintroduction programs is to treat animals for parasites before they are released. Although there is an obvious benefit to release healthy individuals, it might be worth considering whether some infectious diseases should be simultaneously reintroduced to a system with their hosts. For instance, wolves are sometimes reintroduced to help them re-establish parts of their former range where they were extirpated by hunting. Reintroduced wolves are dewormed before release (such as into Yellowstone National Park in 1995). The wolves then enter a habitat potentially free from the tapeworm that debilitates their prey. Might the wolf's reintroduction be more successful with the tapeworm than without it? An important consideration for this example is that the tapeworm is not specific to wolves and, furthermore, can be pathogenic to livestock and humans, which can be accidental hosts. Another consideration is whether to have corridors between reserves (to maintain gene flow and metapopulation dynamics), because this will also allow the spread of infectious diseases among reserves (Hess 1994). If native infectious diseases are a natural part of ecosystems, conservation biologists should not try to exclude them. On the other hand, if introduced diseases are a known threat, isolation and prevention might be a valuable strategy.

Usually, the threats from infectious disease derive from human encroachment into natural habitats (Cleaveland 2009). Prohibiting the arrival of new diseases might be the best way to minimize the effect of infectious diseases on species of concern. However, this is not easy. Lessons from introduced avian malaria in Hawaii did not prevent the arrival of malaria vectors in the Galápagos (Wikelski et al 2004) and the subsequent exposure of Galápagos penguins (*Spheniscus mendiculus*) to malaria (Levin et al 2009). People move animals intentionally all the time and the legal and illegal pet trade ship a baffling number of wild caught animals, including their infectious diseases, to every potential market on earth (Rosen and Smith 2010). Dogs and cats are popular pets, including for people living in and near nature reserves. In poor countries, veterinary care is a low priority, leading to a high prevalence of disease in pets, which are often loose and intermix with wildlife. Even in wealthy countries, pets have several infectious diseases that can be transmitted to wildlife. Though veterinary services are available and laws might prohibit movement of pets, pet owners are often reluctant to follow protocols if it inconveniences them. For instance, on Catalina Island in Los Angeles County, California, pets have exposed the endemic island fox (*Urocyon littoralis*) to canine distemper virus (among other pet-transmitted diseases) (Clifford et al 2006), leading to a near extirpation. Therefore, although quarantines and importation bans are essential tools for protecting wildlife, they are difficult to enact and enforce.

If an epizootic occurs or seems pending, reducing the abundance of susceptible hosts will decrease the chance of disease spread. Vaccination and culling are two ways to reduce the density of susceptible hosts. Vaccination is preferred for endangered species because it protects existing populations and decreases the relative as well as the absolute abundance of susceptible hosts, making it effective against density-dependent and frequency-dependent transmission. The existence of vaccinations for diseases of humans and domestic animals makes it possible to vaccinate endangered wildlife against common viral pathogens. One of the earliest interventions to protect endangered species against infectious diseases was the vaccination of chimps in Gombe against polio (Van-Lawick-Goodall 1971). Vaccination campaigns for several endangered species have been mounted against canine distemper virus and feline leukemia virus, though it is difficult to evaluate success, particularly without unvaccinated control populations (Cleaveland 2009). Vaccination programs can be controversial. The campaigns are expensive and require capturing a large proportion of the target population, with attendant risks to animals during handling. Culling is sometimes suggested as a potential option when no vaccine is available. However, to be effective in eradicating an infectious disease, culling often must be severe and persistent. This might be acceptable when the host is a domestic or otherwise common animal (Ferguson et al 2001), but culling a threatened species might put it at greater risk of extinction. Moderate levels of culling were not able to stop the spread of Tasmanian devil facial tumor disease, and models indicated that the level of culling need to eliminate the disease would place the species in substantial risk of extinction (Beeton and McCallum 2011). Culling or vaccinating threatened species are likely to be used only in desperate situations where managers also have substantial resources and access to the threatened species. Nonetheless, given how frequently viruses from domestic animals threaten wild species, this last option might often be worth the cost.

More and more species now only exist in captivity. In the case where a species is being managed with captive breeding, managers should take extra precautions to prevent disease. Animals should be held in at least two separate locations to provide an insurance against contamination of a facility. Workers should observe high standards of hygiene and try to limit actions that would spread infections among individuals. In addition, animals should be housed away from other organisms that could be a source of infectious diseases. Once animals are ready for reintroduction, veterinarians should check for any infections acquired in captivity before release into the wild. This will help reduce the risk of releasing sick animals, which will have a lower probability of surviving and could be a source of infectious disease that could affect other individuals. Due to past catastrophes, it is now more common for veterinarians to be involved in captive breeding programs. While this is standard practice, it can have unintended consequences. When the last remaining California condors were caught for captive breeding, veterinarians treated them for a host-specific louse. The California Condor is recovering in the wild, but the condor louse is now gone—the only known example of conservation biologists intentionally causing an extinction.

In many countries, a PVA is a legal requirement of threatened and endangered species recovery plans. PVAs are useful for determining the prospects of endangered species, but ignoring disease can decrease their accuracy. PVAs are stochastic models that use measures of vital rates (birth, death) and their variance to estimate the expected time until a population will go extinct. For instance, eventually a run of bad years in computer simulation will drive birth rates below replacement, leading to extirpation. Management then tries to identify which vital rates can be improved to try to extend the expected time to extinction to the distant future. Measured rates of death and reproduction include the effects of infectious diseases, but PVAs assume that these rates are inherent to the species. If infectious diseases are important drivers of vital rates, but PVAs do not treat them as density-dependent processes, the results will give overly optimistic estimates of extinction times (Gerber et al 2005). Therefore, managers should carefully consider how important infectious diseases are in their systems before interpreting recommendations from PVAs.

Little effort has gone into planning conservation around infectious disease. This is in part because conservation biology does not often consider infectious diseases at the population level. It is also because managing infectious diseases is difficult. On the other hand, humans have been attempting and sometimes succeeding in managing infectious diseases in human and livestock populations, suggesting that conservation biologists have good models to follow, vaccination programs being the most obvious. Still, if infectious diseases are natural components of ecosystems, these processes should be allowed to play out, though there might be associated risks to human and livestock health that need to be considered and mitigated.

5.8 Conclusion

Conservation biologists do often not think about parasites, and if they do, they tend not to like them. This abhorrence makes sense because infectious diseases have been associated with the extinction and endangerment of some species. However, for conservation biologists to deal with infectious diseases, it is first necessary to understand other factors, for example, habitat loss and overharvest, that are the overwhelming drivers of endangerment. Furthermore, parasites are pervasive and integral components of all ecosystems. They play important roles as natural enemies on par with top predators. Many of these roles are considered positive from a conservation perspective, from regulating population abundances to maintaining species diversity. A world without parasites would be different, and perhaps not better. Parasites can themselves become endangered along with their hosts. However, it seems unlikely that humans will act to protect parasites from extinction. Parasites are sensitive to environmental changes. It is a surprise to most people that parasites are often sensitive to habitat loss and degradation. Ironically, this makes some parasites positive indicators of ecosystem "health". Though I argue it is important for conservation biologists to think about parasites, there are not many management options for dealing with them, apart from minimizing the movement of domestic and invasive species. I hope that with increasing research on the ecology of parasites, we will have more options for managing them in the future.

References

Aguirre-Macedo ML, Vidal-Martínez VM, Lafferty KD (2011) Trematode communities in snails can indicate impact and recovery from hurricanes in a tropical coastal lagoon. Intern J Parasitol 41:1403–1408

Albon SD, Stien A, Irvine RJ, Langvatn R, Ropstad E, Halvorsen O (2002) The role of parasites in the dynamics of a reindeer population. Proc Roy Soc London Ser B 269:1625–1632

Altizer S, Nunn CL, Lindenfors P (2007) Do threatened hosts have fewer parasites? A comparative study in primates. J Anim Ecol 76:304–314

Amundsen PA, Kristoffersen R (1990) Infection of whitefish (*Coregonus lavaretus* L. s.l.) by *Triaenophorus crassus* Forel (Cestoda: Pseudophyllidea): a case study in parasite control. Can J Zool 68:1187–1192

Anderson RM, May RM (1978) Regulation and stability of host-parasite population interactions. I. Regulatory processes. J Anim Ecol 47:219–247

Beeton N, McCallum H (2011) Models predict that culling is not a feasible strategy to prevent extinction of Tasmanian devils from facial tumour disease. J App Ecol 48:1315–1323

Behrens MD, Lafferty KD (2004) Effects of marine reserves and urchin disease on southern California rocky reef communities. Mar Ecol Progr Ser 279:129–139

Ben-Horin T, Lenihan HS, Lafferty KD (2013) Variable intertidal temperature explains why disease endangers black abalone. Ecology 94:161–168

Byers JE, Altman I, Grosse AM, Huspeni TC, Maerz JC (2011) Using parasitic trematode larvae to quantify an elusive vertebrate host. Conserv Biol 25:85–93

Chen HW, Liu WC, Davis AJ, Jordan F, Hwang MJ, Shao KT (2008) Network position of hosts in food webs and their parasite diversity. Oikos 117:1847–1855

Clay K, Reinhart K, Rudgers J, Tintjer Tammy, Koslow J, Flory SL (2008) Red queen communities. In: Ostfeld R, Keesing F, Eviner V (eds) Infectious disease ecology: effects of disease on ecosystems and of ecosystems on disease. Institute for Ecosystem Studies, Millbrook, pp 145–178

Cleaveland S (2009) Viral threats and vaccination: disease management of endangered species. Anim Conserv 12:187–189

Clifford D, Mazet J, Dubovi E, Garcelon D, Coonan T, Conrad P, Munson L (2006) Pathogen exposure in endangered island fox (*Urocyon littoralis*) populations: implications for conservation management. Biolog Conserv 131:230–243

Cunningham AA, Daszak P (1998) Extinction of a species of land snail due to infection with a microsporidian parasite. Conserv Biol 12:1139–1141

de Castro F, Bolker B (2005) Mechanisms of disease-induced extinction. Ecol Lett 8:117–126

Dobson AP (1988) The population biology of parasite-induced changes in host behavior. Quart Rev Biol 63:139–165

Dobson AP, Lafferty KD, Kuris AM, Hechinger RF, Jetz W (2008) Homage to Linnaeus: how many parasites? How many hosts? Proc Nat Acad Sci USA 105:11482–11489

Dobson AP, May RM (1987) The effects of parasites on fish populations—theoretical aspects. Internat J Parasitol 17:363–370

Dunn RR (2002) On parasites lost-and found: passenger pigeon lice rediscovered. Wild Earth 12:28–31

Dunn RR, Harris NC, Colwell RK, Koh LP, Sodhi NS (2009) The sixth mass coextinction: are most endangered species parasites and mutualists? Proc Roy Soc B Biolog Sci 276:3037–3045

Fenner F, Ratcliffe FN (1965) Myxomatosis. Cambridge University Press, Cambridge

Ferguson NM, Donnelly CA, Anderson RM (2001) Transmission intensity and impact of control policies on the foot and mouth epidemic in Great Britain. Nature 413:542–548

Gerber LR, Lafferty KD, McCallum HI, Sabo JL, Dobson AP (2005) Exposing extinction risk analysis to pathogens: is disease just another form of density dependence? Ecolog Appl 15:1402–1414

Hadeler KP, Freedman HI (1989) Predator-prey populations with parasitic infection. J Math Biol 27:609–631

Hechinger RF, Lafferty KD (2005) Host diversity begets parasite diversity: bird final hosts and trematodes in snail intermediate hosts. Proc Roy Soc London Ser B Biolog Sci 272:1059–1066

Hechinger RF, Lafferty KD, Dobson AP, Brown JH, Kuris AM (2011a) A common scaling rule for abundance, energetics, and production of parasitic and free-living species. Science 333:445–448

Hechinger RF, Lafferty KD, McLaughlin JP, Fredensborg BL, Huspeni TC, Lorda J, Sandhu PK, Shaw JC, Torchin ME, Whitney KL, Kuris AM (2011b) Food webs including parasites, biomass, body sizes, and life stages for three California/Baja California estuaries. Ecology 92:791–792

Hess GR (1994) Conservation corridors and contagious disease: a cautionary note. Conserv Biol 8:256–262

Hudson PJ, Dobson AP, Lafferty KD (2006) Parasites and ecological systems: is a healthy system one with many parasites? Trends Ecol Evol 21:381–385

Huspeni TC, Hechinger RF, Lafferty KD (2005) Trematode parasites as estuarine indicators: opportunities, applications and comparisons with conventional community approaches. In: Bortone SA (ed) Estuarine indicators. CRC Press, Boca Raton, pp 297–314

Huspeni TC, Lafferty KD (2004) Using larval trematodes that parasitize snails to evaluate a saltmarsh restoration project. Ecol Appl 14:795–804

Johnson PTJ, Dobson A, Lafferty KD, Marcogliese DJ, Memmott J, Orlofske SA, Poulin R, Thieltges DW (2010) When parasites become prey: ecological and epidemiological significance of eating parasites. Trends Ecol Evol 25:362–371

Keesing F, Belden LK, Daszak P, Dobson A, Harvell CD, Holt RD, Hudson P, Jolles A, Jones KE, Mitchell CE, Myers SS, Bogich T, Ostfeld RS (2010) Impacts of biodiversity on the emergence and transmission of infectious diseases. Nature 468:647–652

Keesing F, Holt RD, Ostfeld RS (2006) Effects of species diversity on disease risk. Ecol Lett 9:485–498

Keymer AE (1982) Density-dependent mechanisms in the regulation of intestinal helminth populations. Parasitology 84:573–587

Koh LP, Dunn RR, Sodhi NS, Colwell RK, Proctor HC, Smith VS (2004) Species coextinctions and the biodiversity crisis. Science 305:1632–1634

Kuris AM, Hechinger RF, Shaw JC, Whitney KL, Aguirre-Macedo L, Boch CA, Dobson AP, Dunham EJ, Fredensborg BL, Huspeni TC, Lorda J, Mababa L, Mancini F, Mora AB,

Pickering M, Talhouk NL, Torchin ME, Lafferty KD (2008) Ecosystem energetic implications of parasite and free-living biomass in three estuaries. Nature 454:515–518
Kutz SJ, Hoberg EP, Polley L, Jenkins EJ (2005) Global warming is changing the dynamics of Arctic host-parasite systems. Proc Roy Soc B 272:2571–2576
Lafferty KD (1992) Foraging on prey that are modified by parasites. Am Nat 140:854–867
Lafferty KD (1997) Environmental parasitology: what can parasites tell us about human impacts on the environment? Parasitol Tod 13:251–255
Lafferty KD (1999) The evolution of trophic transmission. Parasitol Tod 15:111–115
Lafferty KD (2004) Fishing for lobsters indirectly increases epidemics in sea urchins. Ecol Appl 14:1566–1573
Lafferty KD (2009) The ecology of climate change and infectious diseases. Ecology 90:888–900
Lafferty KD (2012) Biodiversity loss decreases parasite diversity: theory and patterns. Philos Trans Roy Soc Biol Sci 367:2814–2827
Lafferty KD, Allesina S, Arim M, Briggs CJ, DeLeo G, Dobson AP, Dunne JA, Johnson PT, Kuris AM, Marcogliese DJ, Martinez ND, Memmott J, Marquet PA, McLaughlin JP, Mordecai EA, Pascual M, Poulin R, Thieltges DW (2008a) Parasites in food webs: the ultimate missing links. Ecol Lett 11:533–546
Lafferty KD, Dobson AP, Kuris AM (2006) Parasites dominate food web links. Proc Nat Acad Sci USA 103:11211–11216
Lafferty KD, Gerber LR (2002) Good medicine for conservation biology: the intersection of epidemiology and conservation theory. Conserv Biol 16:593–604
Lafferty KD, Holt RD (2003) How should environmental stress affect the population dynamics of disease? Ecol Lett 6:797–802
Lafferty KD, Kuris AM (2009) Parasites reduce food web robustness because they are sensitive to secondary extinction as illustrated by an invasive estuarine snail. Philos Trans Roy Soc Biol Sci 364:1659–1663
Lafferty KD, Porter JW, Ford SE (2004) Are diseases increasing in the ocean? Ann Rev Ecol Evol Syst 35:31–54
Lafferty KD, Shaw JC, Kuris AM (2008b) Reef fishes have higher parasite richness at unfished Palmyra Atoll compared to fished Kiritimati Island. Eco Health 5:338–345
Levin I, Outlaw D, Vargas F, Parker P (2009) Plasmodium blood parasite found in endangered Galapagos penguins (*Spheniscus mendiculus*). Biol Conserv 142:3191–3195
Lloyd-Smith JO, Cross PC, Briggs CJ, Daugherty M, Getz WM, Latto J, Sanchez MS, Smith AB, Swei A (2005) Should we expect population thresholds for wildlife disease? Trends Ecol Evol 20:511–519
Longcore J, Pessier A, Nichols D (1999) *Batrachochytrium dendrobatidis* gen. et sp. nov., a chytrid pathogenic to amphibians. Mycology 91:219–227
Marcogliese DJ (2005) Parasites of the superorganism: are they indicators of ecosystem health? Internat J Parasitol 35:705–716
May RM, Anderson RM (1978) Regulation and stability of host-parasite population interactions. II. Destabilizing processes. J Anim Ecol 47:249–267
Miura O, Kuris AM, Torchin ME, Hechinger RF, Dunham EJ, Chiba S (2005) Molecular-genetic analyses reveal cryptic species of trematodes in the intertidal gastropod, *Batillaria cumingi* (Crosse). Internat J Parasitol 35:793–801
Nichols E, Gómez A (2011) Conservation education needs more parasites. Biol Conserv 144:937–941
Packer C, Holt RD, Hudson PJ, Lafferty KD, Dobson AP (2003) Keeping the herds healthy and alert: implications of predator control for infectious disease. Ecol Lett 6:797–802
Poulin R, Morand S (1997) Parasite body size distributions: interpreting patterns of skewness. Internat J Parasitol 27:959–964
Poulin R, Morand S (2004) Parasite biodiversity. Smithsonian, Washington, DC
Purvis A, Gittleman JL, Cowlishaw G, Mace GM (2000) Predicting extinction risk in declining species. Proc Roy Soc B Biol Sci 267:1947–1952

Rosen G, Smith K (2010) Summarizing the evidence on the international trade in illegal wildlife. EcoHealth 7:24–32

Rudolf V, Lafferty KD (2011) Stage structure alters how complexity affects stability of ecological networks. Ecol Lett 14:75–79

Salkeld DJ, Padgett KA, Jones JH (2013) A meta-analysis suggesting that the relationship between biodiversity and risk of zoonotic pathogen transmission is idiosyncratic. Ecol Lett 16:679–686

Sato T, Watanabe K, Kanaiwa M, Niizuma Y, Harada Y, Lafferty KD (2011) Nematomorph parasites drive energy flow through a riparian ecosystem. Ecology 91:201–207

Scott ME (1987) Regulation of mouse colony abundance by *Heligmosomoides polygyrus* (Nematoda). Parasitology 95:111–129

Smith KE, Sax DE, Lafferty KD (2006) Evidence for the role of infectious disease in species extinction and endangerment. Conserv Biol 20:1349–1357

Sonnenholzner JI, Lafferty KD, Ladah LB (2011) Food webs and fishing affect parasitism of the sea urchin *Eucidaris galapagensis* in the Galápagos. Ecology 92:2276–2284

Sprent JFA (1992) Parasites lost. Internat J Parasitol 22:139–151

Thomas F, Schmidt-Rhaesa A, Martin G, Manu C, Durand P, Renaud F (2002) Do hairworms (Nematomorpha) manipulate the water seeking behaviour of their terrestrial hosts? J Evolut Biol 15:356–361

Tompkins DM, Begon M (1999) Parasites can regulate wildlife populations. Parasitol Tod 15:311–313

Tompkins DM, Sainsbury AW, Nettleton P, Buxton D, Gurnell J (2002) Parapoxvirus causes a deleterious disease in red squirrels associated with UK population declines. Proc Roy Soc London Ser B Biol Sci 269:529–533

Torchin ME, Byers JE, Huspeni TC (2005) Differential parasitism of native and introduced snails: replacement of a parasite fauna. Biol Invas 7:885–894

Valtonen ET, Marcogliese DJ, Julkunen M (2010) Vertebrate diets derived from trophically transmitted fish parasites in the Bothnian Bay. Oecologia 162:139–152

Van-Lawick-Goodall J (1971) In the shadow of man. William Collins Sons, Glasgow

Vitone ND, Altizer S, Nunn CL (2004) Body size, diet and sociality influence the species richness of parasitic worms in anthropoid primates. Evol Ecol Res 6:183–199

Ward JR, Lafferty KD (2004) The elusive baseline of marine disease: are diseases in ocean ecosystems increasing? Publ Lib Sci Biol 2:542–547

Whiteman NK, Parker PG (2005) Using parasites to infer host population history: a new rationale for parasite conservation. Anim Conserv 8:175–181

Wikelski M, Foufopoulos J, Vargas H, Snell H (2004) Galapagos birds and diseases: invasive pathogens as threats for island species. Ecol Soc 9:5–15

Wilcove DS, Rothstein D, Dubow J, Phillips A, Losos E (1998) Quantifying threats to imperiled species in the United States. Bioscience 48:607–615

Williams ES, Thorne ET, Appel MJG, Belitsky DW (1988) Canine distemper in black-footed ferrets (*Mustela nigripes*) from Wyoming. J Wildl Dis 24:385–398

Wood CL, Lafferty KD, Micheli F (2010) Fishing out marine parasites? Impacts of fishing on rates of parasitism in the ocean. Ecol Lett 13:761–775

Chapter 6
The Conservation Value of Agricultural Landscapes

Luciano M. Verdade, Marli Penteado, Carla Gheler-Costa, Graziella Dotta, Luís Miguel Rosalino, Vânia Regina Pivello, Carlos I. Piña and Maria Carolina Lyra-Jorge

Abstract Agricultural landscapes are formed by a matrix of agricultural fields (i.e., agroecosystems) more or less interspersed by remaining fragments of native vegetation, water courses, roads, and human habitations. Although their presumptive

L. M. Verdade (✉)
Centro de Energia Nuclear na Agricultura, Universidade de São Paulo,
Caixa Postal 96, Piracicaba, SP 13416-000, Brazil
e-mail: lmverdade@usp.br

M. Penteado
Estação Ecológica Tupinambás, ICMBio/Ministério do Meio Ambiente,
Av. Manoel Hyppólito do Rego, 1907, São Sebastião, SP 11600-000, Brazil
e-mail: marli.penteado@icmbio.gov.br

C. Gheler-Costa
Ecologia Aplicada, Universidade Sagrado Coração (USC),
Rua Irmã Arminda, 10–50 Jardim Brasil, Bauru, SP 17011-160, Brazil

G. Dotta
Conservation Science Group, Department of Zoology,
University of Cambridge, Downing Street, Cambridge CB2 3EJ, UK

L. M. Rosalino
Centro de Biologia Ambiental /Faculdade de Ciências da Univrsidade de Lisboa,
Ed. C2, 5º Piso, Campo Grande, 1749-016 Lisboa, Portugal
e-mail: lmrosalino@fc.ul.pt

L. M. Rosalino
Laboratório de Ecologia Isotópica, CENA/Universidade de São Paulo,
Caixa Postal 96, Piracicaba, SP 13416-000, Brazil

V. R. Pivello
Instituto de Biociências (IB), Department of Ecology,
Universidade de São Paulo,Cidade Universitária, São Paulo, SP, Brazil

C. I. Piña
Centro de Investigaciones Científicas y Transferencia de Tecnología a la Producción,
(CICYTTP-CONICET)/FCyT-UAdER/FCAL-UNER, Dr. Materi y España, CP 3105,
Diamante, Entre Ríos, Argentina
e-mail: cidcarlos@infoaire.com.ar

M. C. Lyra-Jorge
Curso de Ciências Biológicas, Universidade de Santo Amaro, Rua Prof. Eneas de Siqueira
Neto, 340, São Paulo, SP 04829-300, Brazil
e-mail: carolina_ecologia@hotmail.com

L. M. Verdade et al. (eds.), *Applied Ecology and Human Dimensions in Biological Conservation*, DOI: 10.1007/978-3-642-54751-5_6,

mission is the production of domesticated species or their commodities, agricultural landscapes always support some wild species, which can demand human efforts to be conserved, controlled, used, or simply monitored. Such efforts can only be effective if public policy recognizes the multifunctionality of agricultural landscapes, which should be based on the following principles: (a) The conservation value of agricultural landscapes is more related to the landscape β-diversity than to the matrix α-diversity; (b) the agricultural impacts on biodiversity transcend the limits of agricultural landscapes affecting water courses and nature reserves outside them; and (c) agriculture depends on ecosystem and evolutionary services provided by biodiversity in order to be sustainable.

6.1 Human Dimensions of Agriculture

When the first of our ancestors harvested the seeds of an ancestor plant of the current wheat in Mesopotamia approximately 13,000 years ago, he or she started a revolution—called agriculture—that changed the face of the Earth and served as a basis for what we call civilization (Bender 1975; Diamond 2002). Such revolution was based on the manipulation of the evolutionary process by the selection for the non-dehiscence of the seed. This simple change in that plant phenology led to its harvest synchronization which, by its turn, allowed food storage and consequent settlement of their collectors (Barker 1985; Gamble 1986). This successful technology soon became widespread; other plants entered the game, and their wastes could be used to feed animal species that provided meat, milk, and leather, which could also be accumulated, thus generating an unprecedented abundance of resources in a world of scarcity. Human population could then grow with more abundant resources. Such richness concentration brought the necessity of storehouses protected by guardians and organized armies (Garlan 1975). By then, the gods acquired a more humane form (Hart 1986), and organized war—not simple disputes among rival bands—became logistically feasible due to the stored food at rearguard (Coblentz 1986; Flinn et al. 2005). Ironically, agriculture allowed humans to create gods at their own resemblance—eventually claiming the contrary—and kill each other on massive but organized ways (most of the times in the name of those gods) (Lawler 2012).

Despite the ubiquitous occurrence of war along human history, settlements grew in number and size as a consequence of agriculture development (Rykwert 1976; Rich and Wallace-Hadrill 1990). An urban culture then emerged from it with the consequent development of philosophy, sciences, and arts, only possible with a certain *ozio creativo*, although such concept has only been credited by De Masi (1995) as a later achievement of post-industrial societies.

The continuous development of agriculture and urban settlements stimulated the trade among different peoples on a positive feedback until agriculture fields and cities became globally widespread. However, such land use change displaced extensive areas of pristine ecosystems (Foley et al. 2005), promoted soil erosion and loss of fertility (Lal 2008), contaminated the water, caused a massive loss of natural resources, and significantly changed the composition of the atmosphere (Karl and Trenberth 2003; Lashot and Ahuja 1990). Ample and profound impacts of such processes affected the biota (Blois et al. 2013; Parmesan 2006; Peñuelas and Fillela 2001; Thomas et al. 2004). However, paradoxically, agricultural landscapes are a valuable heritage for many cultures for their history, geography, and even for their biodiversity as a relevant part of their memory (Schama 1995).

6.2 Biological Dimensions of Agriculture

The first biological impacts of agriculture on wild species were the domestication process primarily of useful plants and animals and the battle against the so-called plagues (O'Rourke 2000; Rival 1998). In fact, the former has been actively domesticated, whereas the later has become domestic and more resistant over time (Descola 1987). The manipulation of the evolutionary process for utilitarian purposes has resulted in the genetic modification of species phenology, reproductive biology, behavioral ecology, ecophysiology, and feeding ecology (Price 1984; Trut et al. 2009). Moreover, the transformation of pristine ecosystems into anthropic environments and the increased abundance of domesticated species have resulted in the non-utilitarian selection of many undesirable domestic species besides the obvious rats and cockroaches.

The taxa unable to adapt to human domains perished or became restricted to relicts of pristine ecosystems. For this reason, direct loss of biodiversity due to habitat destruction is possibly the main impact of agriculture on biodiversity in historical terms, although it still occurs in some regions of the world (e.g., Koh et al. 2011). In addition, current land use change in agricultural landscapes can cause secondary impacts on biodiversity by the intensification of agricultural practices on already degraded lands, with consequent contamination of the biota and the physical environment (Ceotto 2008; Hellmann and Verburg 2010; Meche et al. 2009; Schiesari and Grillitsch 2011; Verdade et al. 2012). In such circumstance, the introduction of exotic invasive species tends to increase the extinction rate even more and homogenize fauna and flora in large scale in relatively short term (Magnusson et al. 2006), although this can be seen as a natural process with a possible anthropic raise in biological diversity on a long-term basis (Thomas 2013).

When pristine environments are converted into agricultural landscapes, or when a transformation within the later occurs (e.g., replacement of pastureland by forest plantations), the fate of species that inhabited those primordial environments depends on their "ecological versatility" (MacNally 1995), or their utility to humans. Species unable to acclimate or adapt to the changes and without a clear importance to humans will surely get extinct. For example, Chamberlain and Fuller (2000) showed that the local extinction of 33 % of bird species that already inhabited agriculture landscapes in England and Wales were due to recent changes in agricultural land use. A more impressive example comes from Singapore, where in an area of 540 km^2, 5 % of amphibians and reptiles, 30 % of birds, and 40 % of fish and mammal species have been extinct due to the removal of 95 % of the

Table 6.1 Wild species of fauna and flora found in landscape matrices (i.e., agroecosystems) as resident species

Region	Agroecosystem	Taxa	References
Nepal	Subsistance farming systems	Trees	Acharya (2006)
Iberian Peninsula	Agroforest system	Mammals	Rosalino et al. (2009)
	Eucalyptus globulus plantations	Understory vegetation	Carneiro et al. (2007)
The USA	Switchgrass	Migratory birds	Tolbert and Wright (1998), Tolbert (1998), Tolbert et al. (1997)
	Perennial crops	Fauna	McLaughlin and Walsh (1998)
Mexico	Coffee plantation systems	Trees and epiphytes, mammals, birds, reptiles, amphibians, and arthropods	Moguel and Toledo (1999)
Costa rica	Banana and coconut plantations	Dung beetle and terrestrial mammals	Harvey et al. (2006)
NE Brazil	Cocoa plantations	Bats	Faria et al. (2006)
SE Brazil	*Eucalyptus* spp. plantations	Mammals	Lyra-Jorge et al. (2008), Gheler-Costa et al. (2012)
		Birds	Penteado (2006), Millan (2013)
		Amphibians	Lopes (2010)
	Sugarcane plantations	Mammals	Dotta and Verdade (2007, 2009, 2011), Gheler-Costa et al. (2012)
		Birds	Penteado (2006)
	Exotic grasslands	Mammals	Dotta and Verdade (2007, 2009, 2011), Gheler-Costa et al. (2012)
		Birds	Penteado (2006)

territory forests in the last 183 years for implementation of agriculture, and later urban areas (Brook et al. 2003).

Such patterns of local extinctions associated with agriculture expansion possibly led to the belief that agroecosystems are basically "non-habitat," being used only as passages by the wildlife (e.g., Fahrig 2001, 2007; Jonsen et al. 2001). However, agroecosystems as the landscape matrix may be neither uniformly unsuitable as habitat nor serve as a fully absorbing barrier to the dispersal of species (Kupfer et al. 2006) as different species have different perceptions of landscape structure (With et al. 1997). As a matter of fact, many species of the fauna and flora can be considered as agroecosystem residents including vertebrates and invertebrates, trees and bushes (Table 6.1). In general, in such circumstance, predators tend to use the landscape as a whole, whereas part of their prey can be resident of the matrix, forming simple but effective trophic structures (Dotta and Verdade 2007; Verdade et al. 2011). These patterns and processes make biodiversity of agricultural landscapes merit conservation efforts to enhance the role of protected areas.

6.3 Conservation Dimensions of Agricultural Landscape

The intrinsic mission of agricultural landscapes is the production of food, energy, and goods based on domestic species. However, in order to be sustainable, agriculture depends on wild lineages of agricultural crops for genetic improvement, for instance, to resist to new pathogens or adapt to different levels of soil fertility, water scarcity, or, more recently, to climate change (Tanksley and McCouch 1997). These wild genes depend not only on the conservation of the so-called viable populations (Shaffer 1981) of such lineages, but also more importantly on the maintenance of the evolutionary process itself, which is only possible in situ. For this reason, and based on its own benefit, agricultural landscapes should incorporate a second—but also crucial—mission: the conservation of wild species. However, the inevitable conflict between the production of domesticated species and conservation of wild species is analogous to the tragedy of the commons (Hardin 1968), where a system tends to collapse whenever the individual's advantage (in the short term) is contrary to the group's advantage. Even in private agricultural lands, the wild genes—and, therefore, the evolutionary process—are common, not private. The only way to assure its maintenance is by a public policy in which agricultural landscapes are considered multifunctional (Martinelli et al. 2010). The recent debate about the Brazilian Forest Code (Metzger et al. 2010) is an actual example of such conflict. The intensification of agricultural practices is behind such debate (Tscharntke et al. 2005).

Agricultural landscapes are composed of agricultural crops, water courses, human constructions (e.g., roads and habitations), and frequently some remnants of primary or secondary native vegetation. Agricultural crops can be called "agroecosystems" as they are agricultural ecosystems with distinct nutrient cycles and species composition (both usually artificially maintained). Some authors consider agroecosystems as synonymous of agricultural landscapes (e.g., Altieri 1999; Conway 1985, 1987), but agroecosystems are rather the agricultural landscape matrix (e.g., Fahrig et al. 2011). This distinction is relevant in order to understand the possible effects of intensification of agricultural practices on biodiversity conservation.

In the recent years, two main strategies that try to resolve the conflicts between agriculture and nature conservation have emerged: wildlife-friendly farming (or land sharing) and land sparing. The first tries to reduce on-farm impacts and maintain high densities of species in farmed landscapes by retaining natural habitat patches or by using low-intensity farming methods (Green et al. 2005; Fischer et al. 2008). On the other hand, the land-sparing strategy attempts to spare existing

[redacted]

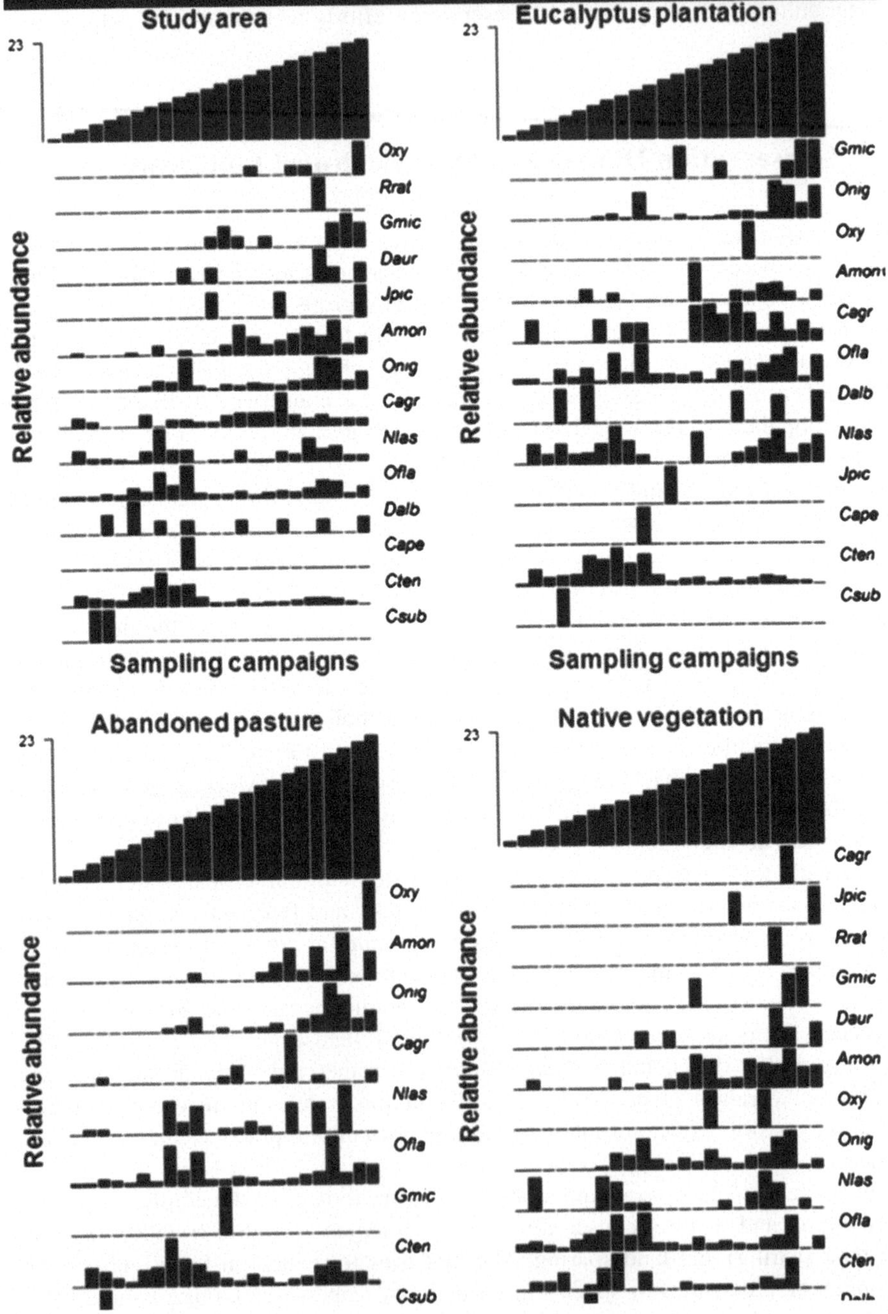
Study area
23
Oxy
Rrat
Gmic
Daur
Jpic
Amon
Onig
Cagr
Nlas
Ofla
Dalb
Cape
Cten
Csub
Relative abundance
Sampling campaigns
Eucalyptus plantation
23
Gmic
Onig
Oxy
Amon
Cagr
Ofla
Dalb
Nlas
Jpic
Cape
Cten
Csub
Relative abundance
Sampling campaigns
Abandoned pasture
23
Oxy
Amon
Onig
Cagr
Nlas
Ofla
Gmic
Cten
Csub
Relative abundance
Native vegetation
23
Cagr
Jpic
Rrat
Gmic
Daur
Amon
Oxy
Onig
Nlas
Ofla
Cten
Relative abundance

← **Fig. 6.1** Temporal variation in the relative abundance of the different small mammal species detected in a silvicultural landscape of Southeastern Brazil (Oxy—*Oxymycterus* spp., Rrat—*Rattus rattus*, Gmic—*Gracilinanus microtarsus*, Daur—*Didelphis aurita*, Jpic—*Juliomys pictipes*, Amon—*Akodon* aff. *montensis*, Onig—*Oligoryzomys nigripes*, Cagr—*Cryptonanus agricolai*, Nlas—*Necromys lasiurus*, Ofla—*O. flavescens*, Dalb—*D. albiventris*, Cape—*Cavia aperea*, Cten—*Calomys tener*, and Csub—*Cerradomys subflavus*). Top graph in each environment represents the cumulative sampling months (from Martin et al. 2012)

natural habitats from future conversion by using high-yielding farming methods in smaller areas (Balmford et al. 2012; Green et al. 2005). The former would be based on agroecosystems, whereas the latter would be based on agricultural landscapes. Both strategies require regulation in order to prevent a tragedy of the commons (Berkes et al. 1989). However, they are still controversial (Vandermeer and Perfecto 2007a; Green et al. 2007).

In general, more attention is paid on the relative reduction in spatial heterogeneity in agroecosystems in relation to natural ecosystems. Such reduction is due to the loss of spatial and biological dimensions as a consequence of the dramatic reduction in species richness from agroecosystems in relation to natural ecosystems, which result in the disappearance of ecological niches (Vandermeer and Perfecto 2007b) and in a simplification of ecological processes (Magnusson 2006). In addition, the expansion of agricultural crops over pristine ecosystems results in their reduction and fragmentation (MacArthur and Wilson 1967; Harris 1984), what increase biodiversity loss even more. However, temporal heterogeneity is considerably larger in agroecosystems than in natural ecosystems. For instance, biomass from a sugarcane field varies from virtually zero to 100 ton ha^{-1} $year^{-1}$ (Goldemberg et al. 2008), whereas *Eucalyptus* plantations vary from bare soil to 300 m^3 ha^{-1} $year^{-1}$ of timber (Stape et al. 2004). Such dramatic variation may lead resident species to deal with famine–feast oscillation in food resources (Wang et al. 2006), with unknown consequences in terms of their demography and evolution. For this reason, for instance, there is a significant variation in resident species composition of small rodents and marsupials along *Eucalyptus* productive commercial cycle in southeastern Brazil (Martin et al. 2012) (Fig. 6.1). In addition, the Montane Akodont (*Akodon montensis*), from *Eucalyptus* plantations, have longer feet than those from surrounding areas of secondary native vegetation (Rosalino et al. 2013), which may be related to adaptive processes even after just a few generations (Linnen et al. 2013). Other species evolved local adaptations to explore new resources available in anthropic landscapes, like vultures (Morán-Lopez et al. 2006), foxes and bears (McDougall et al. 2006), bats, small- and medium-sized terrestrial mammals (Sánchez-Hernandéz et al. 2001), and small mammals (rodents and marsupials) (Tabeni et al. 2005).

Rapid evolution in anthropogenic environments is not exactly a new field of knowledge (e.g., Simpson 1944; Levin 1968; Kettlewell 1955). However, in conservation biology, only recently it is gaining strength (e.g., Ferrière et al. 2004). In agricultural landscapes, it has been traditionally considered in pest control (Palumbi 2001), but future studies of rapid evolution in agricultural landscapes should be prioritized. After all, besides providing biodiversity conservation, agricultural landscapes should provide

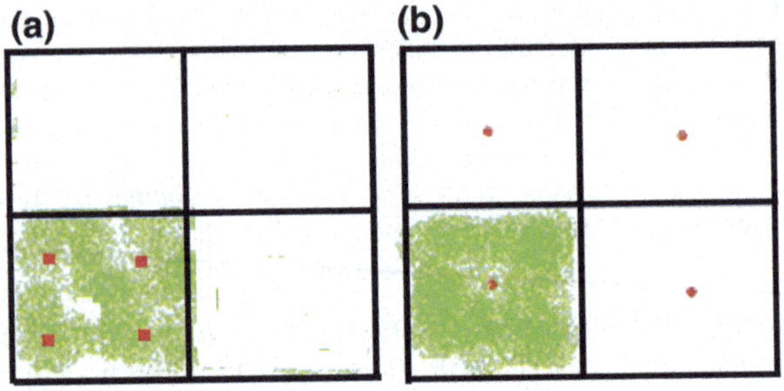

Fig. 6.2 Sketches of biodiversity sampling designs in agricultural landscapes. **a** Sampling concentrated on conservation areas; **b** sampling distributed over the whole agricultural landscape (*Green* remnants of native vegetation; *White* agroecosystem)

the maintenance of the evolutionary process itself as it molds the patterns of biological diversity. Faith and Pollock (2014) call it "evosystem services" in counter-position to the approach of "ecosystem services," usually related to present patterns of abundance and distribution of wild species and their ecological processes. A global long-term ecological and evolutionary research program in agricultural landscapes and pristine ecosystems is necessary to do so (Fig. 6.2). Such program should be based on the following paradigms:

(a) The conservation value of agricultural landscapes is more related to the landscape β-diversity than to the matrix α-diversity (Fig. 6.2);
(b) The agricultural impacts on biodiversity transcend the limits of agricultural landscapes affecting water courses, air composition, and protected areas;
(c) Agriculture depends on the ecosystem and evolutionary services provided by biodiversity in order to be sustainable.

Based on (a) and (b) above, agricultural landscapes should be included in the context of biological conservation, and regulations should be improved. In addition, based on (c) above, biodiversity conservation should be included in the context of agriculture and baselines for its monitoring should be defined (Verdade et al. 2014). Last but not least, it would be possible to establish public environmental and agricultural policies that would assure the multi-functionality of agricultural landscapes. This way, agricultural landscape could complement protected areas in the conservation of biodiversity.

References

Acharya KP (2006) Linking trees on farms with biodiversity conservation in subsistence farming systems in Nepal. Biodivers Conserv 15:631–646

Altieri MA (1999) The ecological role of biodiversity in agroecosystems. Agricult Ecosyst Environ 74:19–31

Balmford A, Green R, Phalan B (2012) What conservationists need to know about farming. Proc R Soc B 279:2714–2724

Barker G (1985) Prehistoric farming in Europe. Cambridge University Press, Cambridge

Bender B (1975) Farming in prehistory: from hunter-gatherer to food producer. John Baker, London. Berkes F, Feeny D, McCay B, Acheson J. 1989. The benefits of the commons. Nature 340:91–93

Berkes F, Feeny D, McCay B, Acheson J (1989) The benefits of the commons. Nature 340:91–93

Blois JL, Zarnetske PL, Fitzpatrick MC, Finnegan S (2013) Climate change and the past, present and future of biotic interactions. Science 342:499–504

Brook BW, Sodhi NS, Ng PKL (2003) Catastrophic extinctions follow deforestation in Singapore. Nature 424:420–426. doi:10.1038/nature01795

Carneiro M, Fabião A, Martins MC, Cerveira C, Santos C, Nogueira C, Lousã M, Hilário L, Fabião A, Abrantes M, Madeira M (2007) Species richness and biomass of understory vegetation in a *Eucalyptus globulus* Labill. coppice as affected by slash management. Eur J Forest Res 126:475–480

Ceotto E (2008) Grassland for bioenergy production. A review. Agron Sust Dev 28:47–55

Chamberlain DE, Fuller RJ (2000) Local extinctions and changes in species richness of lowland farmland birds in England and Wales in relation to recent changes in agricultural land-use. Agric Ecosyst Environ 78:1–17

Coblentz SA (1986) From arrow to atom bomb: the psychological history of war. Peter Owen, London

Conway GR (1985) Agroecosystem analysis. Agric Adm 20(3):1–55

Conway GR (1987) The properties of agroecosystems. Agric Syst 24:95–117

De Masi D (1995) *L'ozio creativo—Conversazione con Maria Serena Palieri*. Roma Ediesse

Descola P (1987) La nature domestique: symbolisme et praxis dans l'écologie des Achuar. Editions de la Maison des Sciences de l'Homme. Man, New Ser 22:754–755

Diamond J (2002) Evolution, consequences and future of plant and animal domestication. Nature 418:700–707

Dotta G, Verdade LM (2007) Trophic categories in a mammal assemblage: diversity in an agricultural landscape. Biota Neotropica 7:287–292

Dotta G, Verdade LM (2009) Felids in an agricultural landscape in São Paulo, Brazil. CATnews 51:22–25

Dotta G, Verdade LM (2011) Medium to large-sized mammals in agricultural landscapes of South-eastern Brazil. Mammalia 75:345–352

Fahrig L (2001) How much habitat is enough? Biol Conserv 100:65–74

Fahrig L (2007) Non-optimal animal movement in human-altered landscapes. Funct Ecol 21:1003–1015

Fahrig L, Baudry J, Brotons L, Burel FG, Crist TO, Fuller RJ, Sirami C, Siriwardena GM, Martin J-L (2011) Functional landscape heterogeneity and animal biodiversity in agricultural landscapes. Ecol Lett 14:101–112

Faith D, Pollock LJ (2014) Phylogenetic diversity and the sustainable use of biodiversity. In: Verdade LM, Lyra-Jorge MC, Piña CI (eds) Applied ecology and human dimensions in biological conservation. Springer, Heidelberg

Faria D, Soares-Santos B, Sampaio E (2006) Bats from the Atlantic rainforest of Southern Bahia, Brazil. Biota Neotropica 6(2):1–13

Ferrière R, Dieckmann Couvet D (2004) Evolutionary conservation biology. Cambridge University Press, Cambridge

Fischer J, Brosi B, Daily GC, Ehrlich PR (2008) Should agricultural policies encourage land sparing or wildlife-friendly farming? Front Ecol Environ 6(7):380–385

Flinn MV, Geary DC, Ward CV (2005) Ecological dominance, social competition, and coalitionary arms races: why humans evolved extraordinary intelligence. Evol Hum Behav 26:10–46

Foley JA, DeFries R, Asner GP, Barford C, Bonan G, Carpenter SR, Chapin FS, Coe MT, Daily GC, Gibbs HK, Helkowski JH, Holloway T, Howard EA, Kucharik CJ, Monfreda C, Patz JA, Prentice IC, Ramankutty N, Snyder PK (2005) Global consequences of land use. Science 309:570–574

Gamble C (1986) The Paleolithic settlement of Europe. Cambridge Univerity Press, Cambridge

Garlan Y (1975) War in ancient world. Chatto & Windus, London
Gheler-Costa C, Vettorazzi CA, Pardini R, Verdade LM (2012) The distribution and abundance of small mammals in agroecosystems of Southeastern Brazil. Mammalia 76:185–191
Goldemberg J, Coelho ST, Guardabassi P (2008) The sustainability of ethanol production from sugarcane. Energ Policy 36:2086–2097
Green RE, Cornell SJ, Scharlemann JPW, Balmford A (2005) Farming and the fate of wild nature. Science 307:550–555
Green RE, Cornell SJ, Scharlemann JPW, Balmford A (2007) Response. Science 308:1257–1258
Hardin G (1968) The tragedy of the commons. Science 162:1243–1248
Harris L (1984) The fragmented forest: island biogeography theory and the preservation of biotic diversity. The University of Chiago Press, Chicago
Hart G (1986) A dictionary of Egyptian gods and goddesses. Routledge & Keagan Paul, London
Harvey CA, Gonzalez J, Somarriba E (2006) Dung beetle and terrestrial mammal diversity in forests, indigenous agroforestry systems and plantain monocultures in Talamanca, Costa Rica. Biodivers Conserv 15:555–585
Hellmann F, Verburg PH (2010) Impact assessment of the European biofuel directive on land use and biodiversity. J Environ Manage 91:1389–1396
Jonsen ID, Bourchier RS, Roland J (2001) The influence of matrix habitat on *Aphthona* flea beetle immigration to leafy spurge patches. Oecologia 127:287–294
Karl TR, Trenberth KE (2003) Modern global climate change. Science 302:1719–1723
Kettlewell HBD (1955) Selection experiments on industrial melanism in the Lepdoptera. Heredity 9:323–342
Koh LP, Miettinen J, Ghazoul J (2011) Remotely sensed evidence of tropical peatland conversion to oil palm. Proc Nat Acad Sci USA 108:5127–5132
Kupfer JA, Malanson GP, Franklin SB (2006) Not seeing the ocean for the islands: the mediating influence of matrix-based processes on forest fragmentation effects. Glob Ecol Biogeogr 15:8–20
Lal R (2008) Soils and sustainable agriculture. A review. Agron Sustain Dev 28:57–64
Lashof DA, Ahuja DR (1990) Relative contributions of greenhouse gas emissions to global warming. Nature 344:529–532
Lawler A (2012) Civilization's double-edged sword. Science 336:832–833
Levin R (1968) Evolution in changing environments. Princeton University Press, Princeton
Linnen CR, Poh Y-P, Peterson BK, Barrett RDH, Larson JG, Jensen JD, Hoekstra HE (2013) Adaptive evolution of multiple traits through multiple mutations at a single gene. Science 339:1312–1316
Lopes PC (2010) Distribuição e abundância de anfíbios e répteis neotropicais em paisagem silvicultural em São Paulo, Brasil. Dissertação de Mestrado. Universidade de São Paulo. Piracicaba. 77 p
Lyra-Jorge MC, Ciocheti G, Pivello VR (2008) Carnivore mammals in a fragmented landscape in northeast of são paulo state, Brazil. Biodivers Conserv 17:1573–1580
MacArthur RH, Wilson EO (1967) The theory of island biogeography. Princeton University Press, Princeton
MacNally RC (1995) Ecological versatility and community ecology. Cambridge University Press, Cambridge
Magnusson WE (2006) Homogeneização biótica. In: Rocha CFD, Bergallo HG, van Sluys M, Alves MAS (eds) Biologia da Conservação: Essências. RiMa Editora, São Carlos, pp 211–229
Martin PS, Gheler-Costa C, Lopes PC, Rosalino LM, Verdade LM (2012) Terrestrial non-volant small mammals in agro-silvicultural landscapes of Southeastern Brazil. For Ecol Manag 282:185–195
Martinelli LA, Joly CA, Nobre CA, Sparovek G (2010) A falsa dicotomia entre a preservação da vegetação natural e a produção agropecuária. Biota Neotrop 10:323–330
McDougall PT, Réale D, Sol D, Reader SM (2006) Wildlife conservation and animal temperament: causes and consequences of evolutionary change for captive, reintroduced, and wild populations. Anim Conserv 9:39–48

McLaughlin S, Walsh M (1998) Evaluating environmental consequences of producing herbaceous crops for bioenergy. Biomass Bioenergy 14:317–324
Meche A, Martins MC, Lofrano BESN, Hardaway CJ, Merchant M, Verdade LM (2009) Determination of heavy metals by inductively coupled plasma-optical emission spectrometry in fish from the Piracicaba River in Southern Brazil. Microchem J 94:171–174
Metzger JP, Lewinsohn TM, Joly CA, Verdade LM, Martinelli LA, Rodrigues RR (2010) Brazilian law: full speed in reverse? Science 329:276–277
Millan CH (2013) Influencia de práticas de manejo e contexto da paisagem sobre a ocorrência de aves em plantio exótico de eucalipto. Dissertação de mestrado. Universidade de São Paulo. São Paulo. 43 p
Moguel P, Toledo VM (1999) Biodiversity conservation in traditional coffee systems of Mexico. Conserv Biol 13:11–21
Morán-López R, Sánchez JM, Costillo E, Corbacho C, Villegas A (2006) Spatial variation in anthropic and natural factors regulating the breeding success of the cinereous vulture (*Aegypius monachus*) in the SW Iberian Peninsula. Biolog Conserv 130:169–182
O'Rourke E (2000) The reintroduction and reinterpretation of the wild. J Agric Environ Ethics 13:145–165
Palumbi SR (2001) Humans as the world's greatest evolutionary force. Science 293:1786–1790
Parmesan C (2006) Ecological and evolutionary responses to climate change. Annu Rev Ecol Evol Syst 37:637–639
Penteado M (2006) Distribuição e abundância de aves em relação ao uso da terra na bacia do Rio Passa-Cinco, estado de São Paulo, Brasil. Tese de Doutorado. Universidade de São Paulo. Piracicaba. 131 p
Peñuelas J, Fillela I (2001) Responses to a warming world. Science 294:793–794
Price EO (1984) Behavioral aspects of animal domestication. Quart Rev Biol 59:1–32
Rich JW, Wallace-Hadrill A (eds) (1990) City and country in ancient world. Routledge, London
Rival L (1998) Domestication as a historical and symbolic process: wild gardens and cultivated forests in the Ecuadorian Amazon. In: Balée W (ed) Advances in historical ecology. Columbia University Press, New York, pp 232–250
Rosalino LM, Rosário J, Santos-Reis M (2009) The role of habitat patches on mammalian diversity in cork oak agroforestry systems. Act Oecolog 35:507–512
Rosalino LM, Martin PS, Gheler-Costa C, Lopes PC, Verdade LM (2013) Allometric relations of Neotropical small rodents (Sigmodontinae) in anthropogenic environments. Zool Sci 30:585–590
Rykwert J (1976) The idea of a town. Faber & Faber, London
Sánchez-Hernández C, Romero-Almaraz ML, Colin-Martínez H, García-Estrada C (2001) Mamíferos de cuatro áreas com diferente grado de alteración en el sureste de México. Act Zoolog Mex 84:35–48
Schama S (1995) Landscape and memory. Harper Perennial, London 652 p
Schiesari L, Grillitsch B (2011) Pesticides meet megadiversity in the expansion of biofuel crops. Front Ecol Environ 9:215–221
Shaffer ML (1981) Minimum population sizes for species conservation. Bioscience 31:131–134
Simpson GG (1944) Tempo and mode in evolution. Columbia University Press, New York
Stape JL, Binkley D, Ryan MG (2004) Eucalyptus production and the supply, use and efficiency of use of water, light and nitrogen across a geographic gradient in Brazil. For Ecol Manag 193:17–31
Tabeni S, Ojeda RA (2005) Ecology of the Monte Desert small mammals in disturbed and undisturbed habitats. J Arid Environ 63:244–255
Tanksley SD, McCouch SR (1997) Seed banks and molecular maps: unlocking genetic potential from the wild. Science 277:1063–1066
Thomas CD, Cameron A, Green RE, Bakkenes M, Beaumont LJ, Collingham YC, Erasmus BFN, Siqueira MF, Grainger A, Hannah L, Hughes L, Huntley B, van Jaarsveld AS, Midgley GF, Miles L, Ortega-Huerta MA, Peterson AT, Phillips OL, Williams SE (2004) Extinction risk from climate change. Nature 427:145–148

Thomas CD (2013) The Anthropocene could raise biological diversity. Nature 502:7
Tolbert VR (1998) Guest editorial. Biomass Bioenergy 14:301–306
Tolbert VR, Wright LL (1998) Environmental enhancement of U.S. biomass crop technologies: research results to date. Biom Bioener 15:93–100
Tolbert VR, Hanowski J, Chrsitian D, Hoffman W, Schiller A, Lindberg J (1997) Changes in bird community composition in response to growth changes in short-rotation woody crop plantings. ORNL/CP 95955 CONF-970856. Oak Ridge National Laboratory, Oak Ridge
Trut L, Oskina I, Kharlamova A (2009) Animal evolution during domestication: the domesticated fox as a model. BioEssays 31:349–360
Tscharntke T, Klein AM, Kruess A, Steffan-Dewenter I, Thies C (2005) Landscape perspectives on agricultural intensification and biodiversity—ecosystem service management. Ecol Lett 8:857–874
Vandermeer J, Perfecto I (2007a) The future of farming and conservation. Science 308:1257
Vandermeer J, Perfecto I (2007b) The agricultural matrix and a future paradigm for conservation. Conserv Biol 21:274–277
Verdade LM, Lyra-Jorge MC, Piña CI (2014) Redirections in conservation biology. In: Verdade LM, Lyra-Jorge MC, Piña CI (eds) Applied ecology and human dimensions in biological conservation. Springer, Heidelberg
Verdade LM, Rosalino LM, Gheler-Costa C, Pedroso NM, Lyra-Jorge MC (2011) Adaptation of mesocarnivores (Mammalia: Carnivora) to agricultural landscapes of Mediterranean Europe and Southeastern Brazil: a trophic perspective. In: Rosalino LM, Gheler-Costa C (eds) Middle-sized carnivores in agricultural landscapes. Nova Science Publishers, New York, pp 1–38
Verdade LM, Gheler-Costa C, Penteado M, Dotta G (2012) The Impacts of sugarcane expansion on wildlife in the state of São Paulo, Brazil. J Sustain Bioener Syst 2:138–144
Wang T, Hung CCY, Randall DJ (2006) The comparative physiology of food deprivation: from feast to famine. Annu Rev Physiol 68:223–251
With KA, Gardner RH, Turner MG (1997) Landscape connectivity and population distributions in heterogeneous environments. Oikos 78:151–169

Part II
Innovation

Chapter 7
The Use of Molecular Tools in Ecological Studies of Mammalian Carnivores

Francisco Palomares and Begoña Adrados

Abstract The use of molecular techniques has been recently incorporated to study the ecology, behavior, and conservation/management of mammalian carnivores, appearing the first paper published only by 1989. The possibility of using non-invasive samples is what increased the use of these techniques. Most of the studies published that use these techniques with mammalian carnivores were conducted during the last 5 years. There are many different molecular techniques available, but not all of them have been commonly used due to the quality of the DNA, which may be degraded and in low quantity in some types of samples such as feces and hairs. The techniques more often used or with a higher future projection are sequencing, PCR, and microsatellites. In most occasions, these techniques have been used to identify species, gender, and individuals, but topics such as landscape ecology, species interactions, foraging ecology, metapopulation dynamics, and conservation medicine might benefit in the near future.

7.1 Introduction

In recent times, molecular tools have strongly burst in the study of individuals, populations, and species. Since the pioneering work by Soulé (1980) brings about the importance of considering population genetic in conservation biology, many

F. Palomares (✉) · B. Adrados
Department of Conservation Biology, Estación Biológica de Doñana,
CSIC, Avda. Américo Vespucio s/n, Isla de la Cartuja, 41092 Sevilla, Spain
e-mail: ffpaloma@ebd.csic.es

B. Adrados
e-mail: b.adrados@ebd.csic.es

L. M. Verdade et al. (eds.), *Applied Ecology and Human Dimensions in Biological Conservation*, DOI: 10.1007/978-3-642-54751-5_7,

researchers have studied different genetic parameters of species or populations. In most occasions, these studies analyzed genetic characteristics of populations or discussed on the potential effects that a reduced genetic variability could have on population persistence (e.g., Mitton and Raphael 1990; Kennedy et al. 1991; Miththapala et al. 1991). But molecular tools are also being used to answer other ecological, behavioral, or specific conservation questions more directly and immediately related to species conservation (e.g., Evans et al. 1989; Packer et al. 1991; Hedrick 1995; Fernández et al. 2006), beyond of only knowing about the genetic structure of populations.

The use of molecular techniques to address ecological issues is known as molecular ecology. It uses DNA markers, which are stable, discrete, and inheritable, to identify species, populations, or individuals and to study the relationships among these (e.g., Höss et al. 1992; Morin and Woodruff 1996; Taberlet et al. 1999). This is a recent incorporation to ecological studies as it is shown by the fact that the now popular scientific journal Molecular Ecology only appeared in 1992.

The aim of this chapter is to highlight the use of molecular tools in the study of ecology, behavior, and conservation/management of mammalian carnivores. Although by briefness in the title and thereafter in the chapter we only write "ecology," we also included in this review studies that used molecular tools to answer behavioral or conservation/management questions. Thus, we do not consider the use of these techniques to know about pure genetic parameters, phylogeography, taxonomy, or even evolution. Even so we recognized the importance of the incorporation of the genetic techniques to advance in the knowledge of these disciplines and their potential application to conservation of species and populations, our aim was to notice the advance that the use of these tools means to answer specific questions in the field of the ecology, behavior, and conservation/management.

Mammalian carnivores are particularly interesting for this field since most of them are nocturnal, have secretive lifestyles, and have low-density populations. Therefore, many ecological questions can not be easily answered with more traditional study methods such as direct observations or trapping. The development of molecular tools for the study of non-invasive samples has specially facilitated the study for this group of animals (e.g., Höss et al. 1992; Foran et al. 1997a; Taberlet and Luikart 1999; Palomares et al. 2002; Pilgrim et al. 2005; Roques et al. 2011).

Our review does not aim to be intensive, rather to be comprehensive about this field. We made a brief summary of (1) the evolution of the use of these techniques during the last decades, (2) the ecological, behavioral, and conservationist/management questions answered using molecular techniques, (3) the main molecular techniques used, and (4) finally some suggestions about the future use of molecular techniques in these fields. Other good reviews related to this topic can be found in Hedrick (2001), Schwartz et al. (2007), Leonard (2008), Segelbacher et al. (2010), and Rodgers and Janečka (2013).

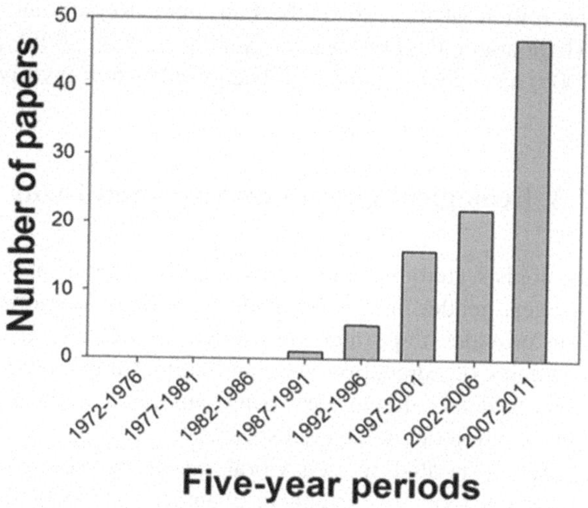

Fig. 7.1 Number of papers published that answered ecological, behavioral, and conservation/management questions using molecular tools in eight major scientific international journals (see text) between 1972 and 2011

7.2 The Use of Molecular Techniques in Ecology

To see how the use of molecular tools has changed along the time to answer ecological questions, we selected and reviewed the number of papers published for this topic using these tools between 1972 and 2011 in eight major international journals old enough (at least published since 1970) to be able to detect these changes during the last decades. The reviewed journals were as follows: Animal Behaviour, Biological Conservation, Ecology, Journal of Animal Ecology, Journal of Applied Ecology, Journal of Mammalogy, Journal of Wildlife Management, and Journal of Zoology.

Results showed that the use of molecular techniques to answer ecological questions is really recent (Fig. 7.1). A total of 90 papers were found, being most of them (51.6 %) published during the last 5 years. Along the time, the increase has been exponential, and in these journals, the first paper using molecular tools to answer an ecological question was published in 1989.

We also conducted a wider review including any journal in order to detect the first use of molecular techniques to answer an ecological question in mammalian carnivores, but we do not detect any before that mentioned in 1989. It is quite recent, after the pioneering papers from Foran et al. (1997a, b) and Taberlet et al. (1997) who developed molecular methods for using with non-invasive samples in several species of carnivores, when the use of molecular tools slowly widespread in ecological studies of mammalian carnivores. The most common non-invasive samples used are feces (Hansen and Jacobsen 1999; Palomares et al. 2002; Verma et al. 2003; Valière et al. 2003; Hedmark et al. 2004; Dalén et al. 2004; Bidlack et al. 2008), which contain many sloughed epithelial cells on their surface, and hair (Valière et al. 2003; Gachot-Neveu et al. 2009), where DNA is extracted from the follicle at the end of the hair shaft. Regurgitates (Valière et al. 2003), urine (Hedmark et al. 2004), saliva (Sundqvist et al. 2008), and blood in snow (Scandura 2005) have occasionally also been used as non-invasive samples.

7.3 Ecological Questions Answered Using Molecular Tools

Ecologists studying carnivores usually employ molecular techniques to identify species, gender, and individuals from these non-invasive samples. This information provides important knowledge for behavioral ecology, species distribution, estimates of population size, and animal movement patterns. It has also applications to detect illegal harvesting, hunt, and trade, which can be essential for the conservation of endangered species.

Topics related to behavioral questions were the first ones answered using molecular tools. For instance, Evans et al. (1989) study the social structure, interrelationship among group members, and paternity in European badgers, *Meles meles*, and found that mating system was either polygynous or promiscuous. A couple of years later, Packer et al. (1991) showed how pride members in African lions, *Panthera leo*, were close relatives in females, but in males, these may be either related or unrelated; in addition, male lion reproductive success becomes skewed as coalition size increased, the males mainly being "helpers" in coalitions composed of close relatives.

More recently, it has been showed using genetic markers that dispersal in a solitary felid species, the bobcat, *Lynx rufus*, is male biased (Janečka et al. 2007) or that the spatial structure (encroachment of their home ranges by neighbors and home-range overlap tolerance) of swift foxes, *Vulpes velox*, or black bears, *Ursus americanus*, was explained by the degree of genetic relatedness among individuals (Kitchen et al. 2005; Moyer et al. 2006).

Diet analysis improved considerably after using molecular techniques to identify the predator and/or prey species. Farrell et al. (2000) could distinguish feces from four sympatric species of carnivores in Venezuela and found that similar sized species as jaguars, *Panthera onca*, and pumas, *Puma concolor*, exhibited minimal prey partitioning, or Shehzad et al. (2012) could differentiate feces from leopard cats, *Prionailurus bengalensis*, and at least 18 different prey species, suggesting the generalist nature of the diet of this felid species.

As mentioned above, after developing molecular tools to be used with non-invasive samples, it was possible to carry out useful studies for the conservation/management of populations of concern species, most of them related to the estimation of population density and abundance. Taberlet et al. (1997) estimated the number and the sex of individuals in a small population of brown bears, *Ursus arctos*. Woods et al. (1999) did it for both black and brown bears, and Ernest et al. (2000) also counted individual pumas, *P. concolor*. Palomares et al. (2002), also using non-invasive samplings of feces that identified with genetic markers, studied the whole distribution area of the most endangered felid in the word, the Iberian lynx, *Lynx pardinus*.

Other conservation/management issues have also recently been able to pose by the use of non-invasive molecular techniques. Blejwas et al. (2006) studied salivary samples obtained from attack wounds on domestic sheep carcasses to identify the species of predator responsible for killings, the sex, and the individual identity; they found that mainly males of coyotes, *Canis latrans*, were responsible for killings. Brøseth et al. (2010), who studied variation in adult survival in a population of wolverine (*Gulo gulo*), found evidence for negative density dependence in survival, which is particularly important for the management of harvest populations.

7.4 Brief Summary of Molecular Techniques Used

There are many different molecular techniques available to be used for answering ecological questions. However, not all of them have been commonly used in ecological studies, neither for carnivores nor for other mammals. The election of the technique should be based on the efficiency to answer the question rather than on the degree of sophistication. The requirement of samples in quantity and good quality is a limiting factor for some techniques, as DNA can be degraded and in low quantity in feces and hairs. The presence of chemical inhibitors, especially in feces, can also be important. Costs of the techniques must be taken into account, especially for long-term monitoring, where a great number of samples will be analyzed. Previous information published about the target species is often essential.

7.4.1 RFLPs

Restriction fragment length polymorphisms (RFLPs) were the first molecular markers that used variations at DNA level, when four decades ago Grodzicker et al. (1974) described this technique for viruses. RFLP consists in detecting polymorphisms though the different patterns of fragments resulted after cutting the DNA with restriction enzymes. The differences in the sequence of DNA of different individuals produce different fragment lengths. The different alleles are the result of deletions, insertions, or even punctual mutations when they occur in the exact enzyme target site (about 4–8 base length). RFLP analyses are slower and more tedious and require larger amounts of DNA than other currently available techniques. It has therefore lost popularity with the constant decreases in prices of sequencing technologies and development of new molecular techniques (see below). Nevertheless, it has been used to identify species in combination with a previous amplification of a region of mtDNA using conserved universal primers (PCR-RFLP) (Paxinos et al. 1997; Hansen and Jacobsen 1999; Rosellini et al. 2008; Bidlack et al. 2008; Mukherjee et al. 2010), which is faster and cheaper than the original RFLP technique.

7.4.2 Sequencing

Determining the order of nucleotides in a DNA fragment is possible since the late 1970s, when different sequencing techniques were developed: "plus and minus" (Sanger and Coulson 1975), "chemical sequencing" (Maxam and Gilbert 1977), and the most popular, "chain-terminating inhibitors" (Sanger et al. 1977). This last technique, also known as "Sanger sequencing," required less use of toxic chemicals and radioactivity and was the method of choice in the following decades. With its automatization in DNA-sequencing instruments based on fluorescence, DNA sequencing became reliable, easy, fast, and cheaper and thus widely available for researchers.

Species identification by sequencing specific fragments of DNA and comparing them with reference sequences is one of the main applications of sequencing, which is known as barcoding. Some efforts have been done to propose the mitochondrial cytochrome c oxidase 1 (COI) gene region as the standard barcode for animals (Hebert et al. 2003), and the Barcode of Life Data Systems (BOLD) was created with that intention (Ratnasingham and Hebert 2007). Nevertheless, segments of other mitochondrial genes have also been sequenced and used as barcodes to identify carnivores. For example, to assign feces to pumas, cytochrome b (cytb) (Farrell et al. 2000; Miotto et al. 2007), 16S (Weckel et al. 2006), and ATP6 (Haag et al. 2009; Chaves et al. 2012) have been successfully sequenced.

Next-generation sequencing (NGS) advances involving whole-genome sequencing and whole-population sequencing (metagenomics) can produce great amounts of sequence data at a low cost (Hudson 2008). Improvements in these new techniques and decreases in costs, in addition to the current expansion of reference databases, can bring a new revolution to the study of ecology (Pompanon et al. 2012). One of these technologies, pyrosequencing, can expand the capabilities of molecular methods for dietary analysis and make it suitable for large-scale diet investigations (King et al. 2008). Pyrosequencing can provide data from individual DNA molecules in complex mixtures using short DNA fragments, therefore allowing the use of feces (Valentini et al. 2009). For example, pyrosequencing has been used to study the diet of fur seals (Deagle et al. 2009) and the leopard cat (Shehzad et al. 2012) through their feces.

7.4.3 Specific Diagnostic PCR

In large-scale monitoring studies with one or a few target species, specific diagnostic amplification is generally a better option than sequencing for barcoding. The fewer steps needed, minimizing the chances of contamination, and the reduced costs facilitate the analysis of large number of samples across broad geographical areas (e.g., Palomares et al. 2002).

The PCR, available since it was discovered in 1985 by Kary Mullis, produces millions of copies of specific segments of DNA using the natural function of the Taq DNA polymerase, a thermostable DNA-copying enzyme. The reaction is very sensitive, allowing the amplification of scarce quantities of DNA. Diagnostic amplification products can be designed to be short to facilitate the application of the method to degraded material, such as feces or hair. Fragments can therefore be visualized in an agarose electrophoresis gel. This technique is fast and inexpensive, and with the adequate primer design, it is very precise and reliable. Specific primers for the identification of several sympatric species can be combined in a rapid classificatory protocol PCR (RCP-PCR) (Dalén et al. 2004; Roques et al. 2011). In addition, in certain large surveying studies, it may be cost-effective using a microarray, which allows screening large number of samples with species-specific primers (Davison et al. 2002; but also see Pfunder et al. 2004).

Gender can also be identified by PCR. Three genes are been typically used in carnivores: SRY, only present in the Y chromosome (e.g., Dallas et al. 2003; Kurose et al. 2005); the zinc-finger region (ZF), present in both X and Y chromosomes, but with a 3-bp deletion in the Y chromosome (Pilgrim et al. 2005; Mucci and Randi 2007; Ralls et al. 2010); and the amelogenin gene, which has a 20-bp deletion on the Y chromosome of felids and ursids and can be used for sex determination within these groups (Pilgrim et al. 2005; Zhang et al. 2008; Poole et al. 2001; Kitamura and Ohnishi 2011).

7.4.4 SSCPs

Single-strand conformational polymorphism (SSCP) is a common method for detecting differences in DNA sequences based on the electrophoretic migration behavior of single-stranded DNA. The differences in mobility depend on the secondary structure of the molecule, which is changed significantly with mutations. It is a simple, inexpensive, and sensitive method for detecting whether DNA fragments are identical in sequence or not. SSCP has been widely used in biomedical research, and some studies have successfully used it in population biology since it was first described by Orita et al. (1989). Oliveira et al. (2010) used SSCPs to identify 16 carnivore species using feces and hair. However, other examples of the use of SSCPs to answer ecological questions in carnivores are not very common. Despite its advantages of codominance and low quantities of DNA template required, it needs highly standardized electrophoretic conditions to obtain reproducible results. In addition, this technique is not always capable of detecting differences in DNA sequences, since two different sequences may have the same electrophoretic mobility.

7.4.5 Microsatellites

Microsatellites, also known as simple sequence repeats (SSRs) or short tandem repeats (STRs), are short repeating sequence of two to six nucleotides (e.g., TC) distributed across the genome. They show variable number of repeats, being these different alleles. They are flanked by single sequences, which can be used to design primers and amplify in a PCR. The presence of these repeated motifs in the genome was discovered in the late 1980s (Litt and Luty 1989; Weber and May 1989; Tautz 1989), and since then, they are among the most used molecular markers.

Microsatellites have been intensively used in ecological research for individual (Waits et al. 2001) or species identification (Ernest et al. 2000), although this last use has been less common because of the limited number of copies of nuclear DNA when compared to mtDNA and the possibility of overlapping alleles (Nauta and Weissing 1996). When microsatellites are used to identify individuals with non-invasive samples, it is important to use techniques to minimize and quantify genotyping errors, such as a multitubes approach (Taberlet et al. 1996) and a multiplex preamplification (Piggott et al. 2004), thus preventing allelic dropout (the preferential amplification of only one of the two alleles in heterozygous individuals) and false alleles (amplification products that can be difficult to distinguish from true alleles).

7.4.6 RAPDs

Randomly amplified polymorphic DNAs (RAPDs) are markers that amplify random segments in a huge number of species. First described by Williams et al. (1990), they are based on the statistical probabilities of finding in the genome complementary sites to the sequence of the primers, which are about 10 base pairs. Polymorphisms are due to changes in the sequence of those sites of alignment, giving a semi-unique resulting pattern. Previous knowledge of the target sequence is not needed, and it is a relatively simple technique that allows analyzing an unlimited number of loci. Although RAPDs can be used in genetic identification of individuals, they have only been used in a few ecological studies (e.g., Ratnayeke et al. 2002; Gachot-Neveu et al. 2009). This is probably due to their dominant character (they are less informative than codominant markers, as they cannot distinguish between heterozygotes and homozygotes for a particular segment), problems in experiment reproducibility, limitations to work with degraded samples, and their lower resolving power when compared to other methods such as microsatellites.

7.5 Concluding Remarks and Future Uses

The use of molecular techniques has only been recently incorporated to the study of ecological issues in mammalian carnivores. Although the first uses were to solve behavioral questions, after the development of their application to non-invasive techniques, most uses were related to determine the presence, abundance, and density of species, topics that are particularly difficult in many carnivore species due to their elusive and low abundance nature. Other topics have been more rarely incorporated although these techniques are being slowly used to understand how carnivores organize in space, disperse, mate, and eat, in addition to more applied questions such as predation on prey of concern. However, there is a promising future for employing molecular techniques to address research questions in other fields such as landscape ecology, species interactions, foraging ecology, metapopulation dynamics, and conservation medicine. In addition, the quick development of molecular techniques and the possibility of using them on non-invasive samples open new possibilities to solve research questions that so far could not be raised in very scarce or endangered species and to plan working hypothesis on large-scale studies.

Acknowledgments The research was carried out under the project CGL2010-16902 of the Spanish Ministry of Science and Innovation.

References

Bidlack AL, Merenlender A, Getz WM (2008) Distribution of nonnative red foxes in East Bay oak woodlands. US For Serv Gen Tech Rep PSW 217:541–548

Blejwas KM, Williams CL, Shin GT, McCullough DR, Jaeger MJ (2006) Salivary DNA evidence convicts breeding male coyotes of killing sheep. J Wildl Manag 70:1087–1093

Brøseth H, Flagstad O, Wärdig C, Johansson M, Ellegren H (2010) Large-scale noninvasive genetic monitoring of wolverines using scats reveals density dependent adult survival. Biol Conserv 143:113–120

Chaves PB, Graeff VG, Lion MB, Oliveira LR, Eizirik E (2012) DNA barcoding meets molecular scatology: short mtDNA sequences for standardized species assignment of carnivore noninvasive samples. Mol Ecol Resour 12:18–35

Dalén L, Götherström A, Angerbjörn A (2004) Identifying species from pieces of faeces. Conserv Genet 5:109–111

Dallas JF, Coxon KE, Sykes T, Chanin PRF, Marshall F, Carss DN, Bacon PJ, Piertney SB, Racey PA (2003) Similar estimates of population genetic composition and sex ratio derived from carcasses and faeces of Eurasian otter *Lutra lutra*. Mol Ecol 12:275–282

Davison A, Birks JDS, Brookes RC, Braithwaite TC, Messenger JE (2002) On the origin of faeces: morphological versus molecular methods for surveying rare carnivores from their scats. J Zool 257:141–143

Deagle BE, Kirkwood R, Jarman SN (2009) Analysis of Australian fur seal diet by pyrosequencing prey DNA in faeces. Mol Ecol 18:2022–2038

Ernest HB, Penedo MCT, May BP, Syvanen M, Boyce WM (2000) Molecular tracking of mountain lions in the Yosemite Valley region in California: genetic analysis using microsatellites and faecal DNA. Mol Ecol 9:433–441

Evans PGH, Macdonald DW, Cheeseman CL (1989) Social structure of the Eurasian badger (*Meles meles*): genetic evidence. J Zool London 218:587–595

Farrell LE, Roman J, Sunquist ME (2000) Dietary separation of sympatric carnivores identified by molecular analysis of scats. Mol Ecol 9:1583–1590

Fernández N, Delibes M, Palomares F (2006) Landscape evaluation in conservation: molecular sampling and habitat modeling for the Iberian lynx. Ecol Appl 16:1037–1049

Foran DR, Crooks KR, Minta SC (1997a) Species identification from scat: an unambiguous genetic method. Wildl Soc Bull 25:835–839

Foran DR, Minta SC, Heinemeyer KS (1997b) DNA-based analysis of hair to identify species and individuals for population research and monitoring. Wildl Soc Bull 25:840–847

Gachot-Neveu H, Lefevre P, Roeder JJ, Henry C, Pulle ML (2009) Genetic detection of sex-biased and age-biased dispersal in a population of wild carnivore, the red fox, *Vulpes vulpes*. Zoo Sci 26:45–152

Grodzicker T, Williams J, Sharp P, Sambrook J (1974) Physical mapping of temperature sensitive mutations. Cold Spring Harbor Symp Quart Biol 39:439–446

Haag T, Santos AS, De Angelo C, Srbek-Araujo AC, Sana DA, Morato RG, Salzano FM, Eizirik E (2009) Development and testing of an optimized method for DNA-based identification of jaguar (*Panthera onca*) and puma (*Puma concolor*) faecal samples for use in ecological and genetic studies. Genetica 136:505–512

Hansen MM, Jacobsen L (1999) Identification of mustelid species: otter (*Lutra lutra*), American mink (*Mustela vison*) and polecat (*Mustela putorius*), by analysis of DNA from faecal samples. J Zool 247:177–181

Hebert PDN, Cywinska A, Ball SL, DeWaard JR (2003) Biological identifications through DNA barcodes. Proc R Soc Lond B Biol Sci 270:313–321

Hedmark E, Flagstad O, Segerström P, Persson J, Landa A, Ellegren H (2004) DNA-based individual and sex identification from wolverine (*Gulo gulo*) faeces and urine. Conserv Genet 5:405–410

Hedrick PW (1995) Gene flow and genetic restoration. The Florida panther as a case study. Conserv Biol 9:996–1007

Hedrick PW (2001) Conservation genetics: where are we now? Trends Ecol Evol 16:629–636

Höss M, Kohn M, Paäbo S, Knauer F, Schröder W (1992) Excrement analysis by PCR. Nature 359:199

Hudson ME (2008) Sequencing breakthroughs for genomic ecology and evolutionary biology. Mol Ecol Resour 8:3–17

Janečka JE, Blankenship TL, Hirth DH, Kilpatrick CW, Tewes ME, Grassman LI (2007) Evidence for male-biased dispersal in bobcats *Lynx rufus* using relatedness analysis. Wildl Biol 13:38–47

Kennedy PK, Kennedy ML, Clarkson PL, Liepins IS (1991) Genetic variability in natural populations of the gray wolf, *Canis lupus*. Can J Zool 69:1183–1188

King RA, Read DS, Traugott M, Symondson WOC (2008) Molecular analysis of predation: a review of best practice for DNA-based approaches. Mol Ecol 17:947–963

Kitamura F, Ohnishi N (2011) Characteristics of Asian black bears stripping bark from coniferous trees. Acta Theriol 56:267–273

Kitchen AM, Gese EM, Waits LP, Karki SM, Schauster ER (2005) Genetic and spatial structure within as a swift fox population. J Anim Ecol 74:1173–1181

Kurose N, Masuda R, Tatara M (2005) Fecal DNA analysis for identifying species and sex of sympatric carnivores: a noninvasive method for conservation on the Tsushima Islands, Japan. J Hered 96:688–697

Leonard JA (2008) Ancient DNA applications for wildlife conservation. Mol Ecol 17:4186–4196

Litt M, Luty JA (1989) A hypervariable microsatellite revealed by in vitro amplification of a dinucleotide repeat within the cardiac muscle actin gene. Am J Hum Genet 44:397–401

Maxam AM, Gilbert W (1977) A new method for sequencing DNA. Proc Nat Acad Sci USA 74:560–564

Miotto RA, Rodrigues FP, Ciocheti G, Galetti PM (2007) Determination of the minimum population size of pumas (*Puma concolor*) through fecal DNA analysis in two protected Cerrado areas in the Brazilian southeast. Biotropica 39:647–654

Miththapala S, Seidensticker J, Phillips LG, Goodrowe KL, Fernando SBU, Forman L, O'Brien SJ (1991) Genetic variation in Sri Lankan Leopards. Zoo Biol 10:139–146

Mitton JB, Raphael MG (1990) Genetic variation in the marten, *Martes americana*. J Mamm 71:195–197

Morin PA, Woodruff DS (1996) Noninvasive genotyping for vertebrate conservation. In: Smith TB, Wayne RK (eds) Molecular genetic approaches in conservation. Oxford University Press, New York, pp 298–313

Moyer MA, McCown JW, Eason TH, Oli MK (2006) Does genetic relatedness influence space use pattern? A test on Florida black bears. J Mamm 87:255–261

Mucci N, Randi E (2007) Sex identification of Eurasian otter (*Lutra lutra*) non-invasive DNA samples using ZFX/ZFY sequences. Conserv Genet 8:1479–1482

Mukherjee S, Ashalakshmi CN, Home C, Ramakrishnan U (2010) An evaluation of the PCR-RFLP technique to aid molecular-based monitoring of felids and canids in India. BMC Res Notes 3:159
Nauta MJ, Weissing FJ (1996) Constraints on allele size at microsatellite loci: implications for genetic differentiation. Genetics 143:1021–1032
Oliveira R, Castro D, Godinho R, Luikart G, Alves PC (2010) Species identification using a small nuclear gene fragment: application to sympatric wild carnivores from South-western Europe. Conserv Genet 11:1023–1032
Orita M, Suzuki Y, Sekiya T, Hayashi K (1989) Rapid and sensitive detection of point mutations and DNA polymorphisms using polymerase chain reaction. Genomics 5:874–879
Packer C, Gilbert DA, Pusey AE, O'Brien SJ (1991) A molecular genetic analysis of kinship and cooperation in African lions. Nature 351:562–565
Palomares F, Godoy JA, Piriz A, O'Brien SJ, Johnson WE (2002) Faecal genetic analysis to determine the presence and distribution of elusive carnivores: design and feasibility for the Iberian lynx. Mol Ecol 11:2171–2182
Paxinos E, McIntosh C, Ralls K, Fleischer R (1997) A noninvasive method for distinguishing among canid species: amplification and enzyme restriction of DNA from dung. Mol Ecol 6:483–486
Pfunder M, Holzgang O, Frey JE (2004) Development of microarray-based diagnostics of voles and shrews for use in biodiversity monitoring studies, and evaluation of mitochondrial cytochrome oxidase I vs. cytochrome b as genetic markers. Mol Ecol 13:1277–1286
Piggott M, Bellemain E, Taberlet P, Taylor A (2004) A multiplex pre-amplification method that significantly improves microsatellite amplification and error rates for faecal DNA in limiting conditions. Conserv Genet 5:417–420
Pilgrim KL, Mckelvey KS, Riddle AE, Schwartz MK (2005) Felid sex identification based on noninvasive genetic samples. Mol Ecol Not 5:60–61
Pompanon F, Deagle BE, Symondson WOC, Brown DS, Jarman SN, Taberlet P (2012) Who is eating what: diet assessment using next generation sequencing. Mol Ecol 21:1931–1950
Poole KG, Mowat G, Fear DA (2001) DNA-based population estimate for grizzly bears in north-eastern British Columbia, Canada. Wildl Biol 7:65–75
Ralls K, Sharma S, Smith DA, Bremner-Harrison S, Cypher BL, Maldonado JE (2010) Changes in kit fox defecation patterns during the reproductive season: implications for noninvasive surveys. J Wildl Manag 74:1457–1462
Ratnasingham S, Hebert PDN (2007) BOLD: the barcode of life data system (http://www.barcodinglife.org). Mol Ecol Not 7:355–364
Ratnayeke S, Tuskan GA, Pelton MR (2002) Genetic relatedness and female spatial organization in a solitary carnivore, the raccoon, *Procyon lotor*. Mol Ecol 11:1115–1124
Rodgers TW, Janečka JE (2013) Applications and techniques for non-invasive faecal genetics research in felid conservation. Eur J Wildl Res 59:1–16. doi:10.1007/s10344-012-0675-6
Roques S, Adrados B, Chavez C, Keller C, Magnusson WE, Palomares F, Godoy JA (2011) Identification of Neotropical felid faeces using RCP-PCR. Mol Ecol Resour 11:171–175
Rosellini S, Osorio E, Ruiz-González A, Piñeiro A, Barja I (2008) Monitoring the small-scale distribution of sympatric European pine martens (*Martes martes*) and stone martens (*Martes foina*): a multievidence approach using faecal DNA analysis and camera-traps. Wildl Res 35:434–440
Sanger F, Coulson AR (1975) A rapid method for determining sequences in DNA by primed synthesis with DNA polymerase. J Mol Biol 94:441–448
Sanger F, Nicklen S, Coulson AR (1977) DNA sequencing with chain-terminating inhibitors. Proc Nat Acad Sci USA 74:5463–5467
Scandura M (2005) Individual sexing and genotyping from blood spots on the snow: a reliable source of DNA for non-invasive genetic surveys. Conserv Genet 6:871–874
Schwartz K, Luikart G, Waples RS (2007) Genetic monitoring as a promising tool for conservation and management. Trends Ecol Evol 22:25–33
Segelbacher G, Cushman SA, Epperson BK, Fortin M-J, Francois O, Hardy OJ, Holderegger R, Taberlet P, Waits LP, Manel S (2010) Applications of landscape genetics in conservation biology: concepts and challenges. Conserv Genet 11:375–385

Shehzad W, Riaz T, Nawaz MA, Miquel C, Poillot C, Shah SA, Pompanon F, Coissac E, Taberlet P (2012) Carnivore diet analysis based on next-generation sequencing: application to the leopard cat (*Prionailurus bengalensis*) in Pakistan. Mol Ecol 21:1951–1965

Soulé ME (1980) Thresholds for survival: maintaining fitness and evolutionary potential. In: Soule ME, Wilcox BA (eds) Conservation biology: an evolutionary-ecological perspective. Sinauer Associates, Inc. Publishers Sunderland, Massachusetts, pp 151–169

Sundqvist A-K, Ellegren H, Vilà C (2008) Wolf or dog? Genetic identification of predators from saliva collected around bite wounds on prey. Conserv Genet 9:1275–1279

Taberlet P, Luikart G (1999) Non-invasive genetic sampling and individual identification. Biol J Linn Soc 68:41–55

Taberlet P, Griffin S, Goossens B, Questiau S, Manceau V, Escaravage N, Waits LP, Bouvet J (1996) Reliable genotyping of samples with very low DNA quantities using PCR. Nucl Acids Res 24:3189–3194

Taberlet P, Camarra J-J, Griffin S, Uhrès E, Hanotte O, Waits LP, Dubois-Paganon C, Burke T, Bouvet J (1997) Noninvasive genetic tracking of the endangered Pyrenean brown bear population. Mol Ecol 6:869–876

Taberlet P, Waits LP, Luikart G (1999) Noninvasive genetic sampling: look before you leap. Trend Ecol Evol 14:323–327

Tautz D (1989) Hypervariability of simple sequences as a general source of polymorphic DNA markers. Nucl Acids Res 17:6463–6471

Valentini A, Miquel C, Nawaz MA, Bellemain E, Coissac E, Pompanon F, Gielly L, Cruaud C, Nascetti G, Wincker P, Swenson JE, Taberlet P (2009) New perspectives in diet analysis based on DNA barcoding and parallel pyrosequencing: the trnL approach. Mol Ecol Resour 9:51–60

Valière N, Fumagalli L, Gielly L, Miquel C, Lequette B, Poulle M-L, Weber J-M, Arlettaz R, Taberlet P (2003) Long-distance wolf recolonization of France and Switzerland inferred from non-invasive genetic sampling over a period of 10 years. Anim Conserv 6:83–92

Verma SK, Prasad K, Nagesh N, Sultana M, Singh L (2003) Was elusive carnivore a panther? DNA typing of faeces reveals the mystery. Forensic Sci Int 137:16–20

Waits LP, Luikart G, Taberlet P (2001) Estimating the probability of identity among genotypes in natural populations: cautions and guidelines. Mol Ecol 10:249–256

Weber JL, May PE (1989) Abundant class of human DNA polymorphisms which can be typed using the polymerase chain reaction. Am J Hum Genet 44:388–396

Weckel M, Giuliano W, Silver S (2006) Jaguar (*Panthera onca*) feeding ecology: distribution of predator and prey through time and space. J Zool 270:25–30

Williams JGK, Kubelik AR, Livak KJ, Rafalsk JA, Tingey SV (1990) DNA polymorphisms amplified by arbitrary primers are useful as genetic markers. Nucl Acid Res 18:6531–6535

Woods JG, Paetkau D, Lewis D, McLellan BN, Proctor M, Strobeck C (1999) Genetic tagging of free-ranging black and brown bears. Wildl Soc Bull 27:616–627

Zhang YG, Janečka JE, Li DQ, Duo HR, Jackson R, Murphy WJ (2008) Population survey and genetic diversity of snow leopards *Panthera uncia* as revealed by fecal DNA. Acta Zool Sin 54:762–766

Chapter 8
The Role of Abundance Estimates in Conservation Decision-Making

James D. Nichols

Abstract Initial discussions about conservation of any species or population tend to include questions about just how many animals there are. Indeed, it is often assumed that abundance estimates are critically important to conservation, to the point where obtaining such estimates is sometimes viewed as a necessary prerequisite for management. At a minimum, this view produces a delay in management, and in the worst case, the monitoring of abundance comes to be equated with conservation. Abundance estimates can be important to conservation, but I believe that development of a clear idea of exactly how they are to be used in the conservation process should precede surveys designed to obtain them. In this chapter, I consider the explicit roles of abundance estimation in conservation, first focusing on the uses of such estimates in conservation programs and then turning to appropriate methods for obtaining those estimates.

8.1 Why Estimate Abundance?

Conservation and management of ecological systems entail making decisions about what actions to take in order to bring about desired consequences. Such consequences are frequently evaluated based on changes in system state variables, where a state variable is a characteristic of the system that reflects system status. When conservation involves single species, for example threatened or endangered species or invasive species, abundance is a state variable of choice. When conservation is focused on animal communities or entire ecosystems, species-specific abundances, or diversity metrics derived from abundances, may still be important state variables reflecting system status. Thus, abundance is a state variable that is relevant for many conservation programs.

J. D. Nichols (✉)
Patuxent Wildlife Research Center, Laurel, MD 20708, USA
e-mail: jnichols@usgs.gov

L. M. Verdade et al. (eds.), *Applied Ecology and Human Dimensions in Biological Conservation*, DOI: 10.1007/978-3-642-54751-5_8,

Monitoring can be an expensive endeavor, requiring that we have a clear idea of how monitoring information is to be used for conservation. Many quantities describing ecological systems can be potentially estimated, and it is important to focus on those quantities that are most relevant to management decisions. The following observation by Platt (1964) is just as relevant to monitoring programs as it is to the selection of experiments: "Biology, with its vast informational detail and complexity, is a 'high-information' field, where years and decades can easily be wasted on the usual type of 'low-information' observations and experiments if one does not think carefully in advance about what the most important and conclusive experiments would be." In order to move away from "low-information observations" and make good decisions about what quantities to monitor, we need to consider the roles of monitoring data in the decision process.

8.1.1 Conservation Programs: Elements

Effective conservation programs include several key elements (see Williams et al. 2002). (1) Clear specification of *objectives* is important to any conservation program. In conservation, objectives frequently involve system state. For example, an objective might be to maintain abundance of an endangered species above some specified level. (2) Conservation also requires a set of possible *actions* that can be taken in an effort to meet objectives. These are specified a priori, and decisions at each decision point entail selecting the appropriate action from this set. (3) *Models* are required in order to project the consequences of the different possible actions. Such projections are essentially compared in order to decide which action is "best" at moving the system in a direction that is desirable, with respect to objectives. (4) A *monitoring* program is useful for several reasons, including provision of an estimate of system state to be used as a starting point for projecting the consequences of management actions. (5) Finally, some kind of *decision algorithm* is used to select a good decision, based on the other elements: objectives, actions, models, and monitoring data. Optimization algorithms are frequently used to solve conservation decision problems.

8.1.2 Conservation Programs: Adaptive Management

Adaptive Management (AM) Motivation. Some conservation programs entail single decisions. For example, a decision about land acquisition in a unique system (by "unique," I mean that information about this decision cannot be obtained from other previous decisions or used for other subsequent decisions) may indeed involve a single decision that does not permit learning that may be useful for other, subsequent decisions. More frequently, however, even a decision about land acquisition in a certain kind of habitat may permit learning that can be useful to future

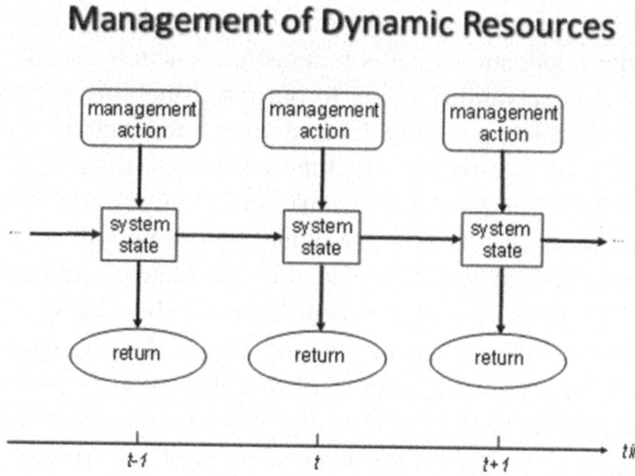

Fig. 8.1 Schematic diagram of a recurrent decision problem in natural resource management

decisions in similar systems. It is very common to manage an animal population or an ecological system associated with a particular piece of land in a manner that requires periodic decisions. For example, we might have to decide what harvest regulations or what habitat management action (e.g., burning) to implement each year. We can refer to such management programs as recurrent decision problems. Such decision problems share two characteristics that are potentially important to their solution, their dynamic nature and the possibility to learn.

A typical recurrent decision problem entails decisions at specified decision points that represent points in time. At any such decision point, *t*, the job of the decision-maker is to select an appropriate action from a set of potential actions. The selected action is taken, producing "rewards" (outcomes that are relevant to stated objectives) and also driving the system to a new state (Fig. 8.1). Because actions are typically state dependent (the appropriate action depends on the current state of the system), and because actions can drive the system to a new state, the decision process should not be solved for a single time step. Instead, the selected action at time *t* should consider not only rewards resulting from the immediate decision, but also the subsequent state of the system and the actions (and rewards) that are appropriate for this state. In fact, for dynamic decision problems, optimal decisions typically consider all future decisions for the entire time horizon of the problem. This complicates the optimization problem, but algorithms such as stochastic dynamic programming (e.g., Bellman 1957; Williams et al. 2002) were developed specifically for this purpose.

Most decision problems in conservation are characterized by substantial uncertainty. This uncertainty can be classified into four categories (Williams 1997): environmental variation, partial controllability, partial observability, and structural uncertainty. Environmental variation and its potential effects on ecological systems are sources of uncertainty that are well known to ecologists and conservation biologists. Partial controllability refers to uncertainty in the application of the management treatment. For example, when fire is used as a tool in habitat management, the amount of area actually burned and the intensity of the burn cannot be precisely controlled by the manager (e.g., Johnson et al. 2011), leading to variation in the application of this management action and the associated system response. Partial observability refers to our usual inability to assess ecological state variables with complete certainty. Instead, quantities such as population size and species richness must be estimated, and the resulting estimates are characterized with sampling variances (see later discussion). Structural uncertainty refers to the fact that we seldom know exactly how the managed system will respond to a specific management action. Instead, we often have multiple plausible hypotheses about system response.

Recurrent decisions permit learning about these sources of uncertainty in order to make better decisions in the future. Learning is primarily focused on structural uncertainty, a source of variation that is potentially reducible (that is, we can potentially learn which hypothesis about system response corresponds most closely to reality). Two other sources of uncertainty, environmental variation and partial controllability, may not be reducible (in the sense that we cannot potentially eliminate them), but recurrent decisions still provide additional information about the form and magnitude of these sources of variation that can be used to aid future decisions. Learning and reducing structural uncertainty are accomplished via an approach that constitutes the key step in the conduct of science. At each decision point, a management action is selected, and competing models about system response to management actions are used to make predictions about system response. Monitoring provides an estimate of system response, and the comparison of this estimate with the different predictions leads to increased faith in the models that predict well and decreased faith in those that predict poorly. This learning, based on embedding this scientific process within the larger decision process, is the hallmark of adaptive resource management (Williams et al. 2007; Nichols and Williams 2013).

The AM process. The AM process includes an initial (and possibly additional) deliberative phase and an iterative phase (see Williams et al. 2007). The initial deliberative phase occurs prior to the initial decision point and the actual implementation of AM. During this initial deliberative phase, objectives and available management actions are specified, models of system response are developed, a monitoring program is established, and a decision algorithm is selected. Armed with these elements, the decision algorithm is used with the current estimate of system state (e.g., population size) to decide what action to take at the first decision point, thus beginning the iterative phase. The action is taken, and the system state changes accordingly. The monitoring program is used to estimate the new system state. This new state is compared against predictions made by the models of system response, and faith is increased in the model(s) that predicts well and decreased for the model(s) that predicted poorly. There is a formal process for this updating of the relative degrees of faith in the different models (see Williams et al. 2002), and this process represents a step in the resolution of structural uncertainty (learning). At the next decision point, the objectives, actions, models (with their updated degrees of faith), and the current state of the system (estimated via monitoring) are then used with the decision algorithm to make the next decision.

The iterative process proceeds in this manner, until there is a reason for revisiting the deliberative phase. For example, as the management process proceeds, perhaps human values change to the point that objectives should be reconsidered. Or perhaps none of the models is predicting very well, leading to reconsideration of the model set. In such cases, the management program can move from the iterative phase to the deliberative phase, a shift referred to as "double-loop learning" (Williams et al. 2007). During this phase, one or more of the decision process elements may be revisited and changes possibly made, and the iterative process is then resumed. The entire process thus consists of both deliberative and iterative phases and is designed to make wise management decisions in the face of uncertainty, in a manner that reduces that uncertainty, thus improving decisions in the future.

8.2 The Role of Abundance Estimates

When AM is applied to animal populations, then abundance estimates serve three critical roles (Yoccoz et al. 2001; Nichols and Williams 2006). The first role involves state-dependent decisions (also see Caughley 1994, 1977). In most cases, the decision of what action to take at any particular decision point will depend on the current state of the system, that is, the abundance of the focal species. If population size is below that specified in objectives as desirable, then wise actions will likely be those that promote population growth. On the other hand, if population size exceeds desired levels, then different actions, or possibly no action, may be called for. The second role of abundance estimates in AM is to assess the degree to which objectives are being met. In conservation settings, desired abundance is frequently an important component of objectives. Abundance estimates thus permit assessment of the effectiveness of management.

The third role of abundance estimates is for use in the learning step of the iterative phase of AM. At each decision point, an action is taken and predictions are made by each of the competing models about how the population will respond. The abundance estimate then provides an indication of actual system response, with which these predictions can be compared. This comparison leads to changes in the degrees of faith associated with the different models, with increased faith for models that predicted well and decreased faith in models that predicted poorly. These updated degrees of faith then determine the influence of the different models in the next decision. Notice that this approach of evaluating hypotheses by comparing predictions of their corresponding models is precisely how science is conducted (e.g., see Williams et al. 2002). Indeed, this step of AM essentially embeds a scientific process within the larger management process. Abundance estimates are critical to this step.

Knowledge of these specific roles of abundance estimates is useful in designing surveys and monitoring programs to inform conservation decisions. For example, the area surveyed will be determined by the area to which conservation actions are to be applied. The timing of the survey will be determined jointly by the timing of the decision process and the biology of the focal species. There is usually an advantage to obtaining abundance estimates very shortly before the time at which the decision is made and the action taken. However, if the focal species is only detectable via the selected survey method at certain times of the year (e.g., breeding season detections of singing male birds), then this kind of consideration may take precedence over basic considerations of the decision process.

8.2.1 What to Estimate

Ecological monitoring programs usually focus on state variables, and the specific state variable selected should be dictated by the larger program of science or conservation that the monitoring is designed to serve. Commonly selected state variables for ecological monitoring programs include abundance, occupancy (the proportion of sites occupied by a species), and species richness. Occupancy and species richness involve abundance to the extent that they focus on whether a species abundance is 0 or >0. Some monitoring programs that focus on communities favor species diversity metrics that include abundance estimates for multiple focal species of the community. The key point is that the selection of a state variable to estimate is inherited directly from the larger program of science or management.

Abundance estimates typically (but not always) require more sampling effort than do occupancy or species richness estimates. It is not uncommon for programs to estimate focal species abundance within a small number of specified areas and to use occupancy modeling over a much larger area (e.g., Karanth et al. 2011). Such an approach provides a picture of species distribution over a perhaps large area of interest and estimates of abundance for selected locations within the area. It is conceptually possible to link these sets of estimates in a way that permits inference about the distribution of abundance across the area of interest (e.g., see Royle and Nichols 2003). In any case, logistical issues such as extent of the area of interest and required survey effort may affect decisions about what state variables to select, but the overriding consideration is what state variables are needed to meet the requirements of the larger program.

8.2.2 How to Estimate Abundance

8.2.2.1 Conceptual Framework

Estimation of animal abundance requires consideration of two important issues: Geographic variation and detectability. Geographic variation refers to the fact that areas for which abundance estimates are needed may be so large as to preclude the possibility of surveying the entire area directly. In some cases, the area of inference is sufficiently small that it can be surveyed in its entirety, and in such cases, geographic variation is not a real issue. In other cases, when interest is in inference about large areas, a subset of sample units is selected to be surveyed, and extrapolation is used to draw inferences about the units not surveyed. This approach requires that sample units be selected for survey in a manner that permits the determination of the probability of being sampled for each potential sampling unit in the population of interest. This information is required to draw inferences about the locations not surveyed, based on those that are surveyed. Various approaches to the selection of sample units may be used including simple random sampling, stratified random sampling, cluster sampling, double sampling, and adaptive sampling (Thompson 2002). This source of variation is one that statisticians have long dealt with in virtually all sample survey problems. Thus, information for dealing with this issue is widely available in the statistical literature (e.g., Cochran 1953; Thompson 2002).

In contrast, the issue of detection probability was historically developed primarily by statisticians and biologists working on animal populations (see reviews in Seber 1982; White et al. 1982). Detectability refers to the fact that even when an animal of a focal species is present in a sample unit that is surveyed, there is some probability that it will be missed in survey efforts. Abundance estimation can be based on many different kinds of count statistics: birds heard, animals caught in traps, ungulates seen from an airplane, animals taken during sport or commercial harvesting, etc. Regardless of the exact nature of the counts, they should be related to abundance, the true quantity of interest, by

$$E(C) = Np, \tag{8.1}$$

where C represents the number of animals counted, N represents the number of animals in the sample unit exposed to sampling efforts, and p is the detection probability, the probability that a member of N appears in the count. $E(C)$ denotes the expected value of the random variable, C. If we could somehow repeat the survey count under the exact same conditions a large number of times, then the mean of these counts would estimate the expectation.

The important point of expression (8.1) is that the count is influenced by two components, one of ecological and conservation interest (N) and one that reflects sampling (p). Thus, in the absence of other information, the data that we collect (C) are not necessarily valuable for inference about abundance. However, if we are somehow able to estimate detection probability, then we can translate our count into an estimate of abundance as follows:

$$\hat{N} = C/\hat{p}, \tag{8.2}$$

where the hats denote estimators. Abundance estimation thus involves obtaining the count and estimating its corresponding detection probability.

8.2.2.2 Indices

Sometimes, biologists and conservationists express more interest in relative abundance than in absolute abundance. For example, define $\lambda_{ij} = N_j/N_i$ as relative abundance, where i and j represent some dimension of comparison. For example, if

$i = t$ represents one year and $j = t + 1$ represents the next year, then λ_{ij} represents the rate of change in abundance or trend. If i and j represent two different locations, then λ_{ij} is the relative abundance of the focal species at the two locations. Or if i and j are two different species, then λ_{ij} represents the relative abundance of these species. Estimation of detection probability requires some effort, so it is tempting to view the count statistics themselves as indices and to use them directly to estimate relative abundance. Indeed, proponents of the use of indices (e.g., Johnson 2008) frequently recommend estimation of λ_{ij} as follows:

$$\hat{\lambda}_{ij}^{C} = C_j / C_i, \tag{8.3}$$

where $\hat{\lambda}_{ij}^{C}$ denotes the estimator of relative abundance that is based on a ratio of counts. In order to evaluate this estimator, we can approximate its expected value as follows (e.g., Williams et al. 2002):

$$E\left(\hat{\lambda}_{ij}^{C}\right) \approx \frac{E(C_j)}{E(C_i)} = \frac{N_j p_j}{N_i p_i} = \lambda_{ij}\left(\frac{p_j}{p_i}\right). \tag{8.4}$$

So the expectation of $\hat{\lambda}_{ij}^{C}$ includes the true parameter of interest, λ_{ij}, but it also includes the ratio of detection probabilities for the two entities (times, places, species) being compared (term in parentheses in right-hand side of Eq. 8.4). If this ratio of detection probabilities is near 1, that is if the two detection probabilities are very similar, then the count-based estimator of (3) may do a good job of estimating the quantity of interest. But if the detection probabilities are dissimilar, then $\hat{\lambda}_{ij}^{C}$ can be a poor estimator of relative abundance, as it confounds true relative abundance with the difference in detection probabilities.

For most dimensions of comparison (locations, species, and frequently time), there will be good reason to expect basic differences in detection probabilities that preclude reasonable use of count-based indices (e.g., see Pinto et al. 2006). When interest is focused on time trend of abundance at specific locations, then it is sometimes argued that even though detection probabilities may vary from year to year, they do some randomly (e.g., they represent random selections from the same statistical distribution year after year), in which case, the estimator of Eq. (8.3) may still perform adequately. That is, on average, the ratio of year-specific detection probabilities will be about 1. However, depending on the kinds of survey methods being used, there are many potential sources of variation that would be expected to cause non-random changes in detection probabilities over time [e.g., shifts in breeding phenology, and thus time-specific calling frequencies, of breeding birds (Crick et al. 1997; Crick 2004); increases in human-generated noise levels over time, potentially influencing auditory surveys].

These considerations lead me to the conclusion that, whenever possible, it is best to collect the ancillary data needed to estimate detection probabilities or to incorporate them directly into modeling efforts (also see Lancia et al. 1994, 2005; Pollock et al. 2002; Williams et al. 2002). Such data permit formal tests for variation in detection probabilities. When such variation does not exist (i.e., when detection probabilities over the dimension of comparison are similar), then this inference can lead to more efficient estimation of λ_{ij}. When evidence of variation in detection does exist, then the ancillary data on detection serve as insurance, permitting inference about relative abundance even in the face of sampling differences.

8.2.2.3 Approaches to Abundance Estimation

As emphasized above, abundance estimation requires some sort of count and an estimate of detection probability that accompanies that count (Eq. 8.2). A large number of methods have been developed for the estimation of animal abundance, filling books (e.g., Seber 1982; Borchers et al. 2002; Williams et al. 2002), and substantive reviews (e.g., Lancia et al. 1994, 2005). These methods entail various count statistics and various corresponding approaches to inference about detection. However, the final step in virtually all of these various methods uses Eq. (8.2), in which a count is divided by the estimated detection probability. Given this variety of methods, how do we decide what method to select for use in conservation? This decision should be based on the larger conservation problem, on how estimates are to be used in the conservation program, on the conceptual framework provided above, and on a variety of logistical issues. These latter issues include such considerations as for what specific areas are abundance estimates needed; how easy or difficult is human travel in these areas; can the focal species be readily detected by sight or sound, or are organisms secretive; and what financial and human resources are available? In summary, selection of appropriate survey and abundance estimation methods should be tailored to the conservation program that those methods are designed to serve, with important considerations being the specific roles of estimates in the program and the logistical issues that accompany the program. It is beyond the scope of this chapter to describe all of the existing approaches to inference about animal abundance. Instead, I will attempt an abbreviated and selective review with pointers to the more detailed literature for readers who desire more information. Abundance estimation methods can be classified in various ways, and here, I will focus on methods that are based on direct observations of unmarked animals and other methods that rely on the ability to identify (usually marked) individuals at multiple points in time.

Direct observations. One of the most widely used methods for abundance estimation is based on the concept of distance sampling (e.g., see Buckland et al. 2001, 2004). Animals are detected via sight or sound by investigators who either traverse a line transect or are stationary at single points. Animals are counted directly, and the ancillary data collected are the estimated distances to each of the detections. If space is sampled randomly, then the distribution of detection distances provides information about detection probabilities, under the reasonable assumption of monotonic decreases in detection probability with distance from the observer. This basic approach has been used with taxa and sampling situations as diverse as avian point counts in forests, line transect surveys of ungulates in grasslands or forest, line transect aerial surveys of organisms ranging from birds to ungulates, and even line transect boat surveys of marine mammals.

Another approach involves multiple observers (e.g., Cook and Jacobson 1979; Nichols et al. 2000; Williams et al. 2002) who survey the same locations at the same times. In addition to the raw counts of the investigators, extra information includes the identity of the investigators who detected each individual organism. So for a two-observer study, some animals will be detected by both observers, some by just one and some by just the other. These three statistics then provide the information to estimate detection probability for each observer and thus the probability that an animal will have been missed by both observers. Survey logistics vary depending on whether the investigator counts are viewed as dependent or independent of each other.

Both distance sampling and multiple observers estimate the number of organisms that are potentially available for detection at the time of the survey. If some animals are in the area being surveyed, but not available for detection during the survey period (e.g., a marine mammal that is submerged for the survey period), then such animals will not be included in the abundance estimate. Time-of-detection models (Farnsworth et al. 2002) were developed to deal with the availability issue for surveys in which individual animals might be unavailable for some part of the survey period and then available for another part of the period. The approach was developed for point-count surveys of vocalizing birds, in which a bird may not vocalize even once during the entire period of a count. The period of the survey at each point can be divided into discrete sampling periods. For example, a 10-min count might be divided into the initial 5 min and the subsequent 5 min. Using one approach to such surveys, an observer records which individuals were detected in the initial 5 min and which individuals were not detected during the initial 5 min but were detected in the second 5-min period. This approach permits estimation of detection probability that includes both the probability that an individual bird is available (potentially detectable) and the probability that it is detected, given that it is available at some time during the survey. Resulting abundance estimates include animals that were not available for detection at any time during the survey, as well as those animals that were available but went undetected (see Nichols et al. 2009).

Another recent approach to abundance estimation is the *N*-mixture models of Royle (2004) that are based on repeat counts at a survey location. If the counts are conducted in rapid succession (e.g., count of total birds heard for 5 min, followed by another count for the next 5 min), then, as with the time-of-detection approach, the abundance estimates should be estimating the number of birds in the sampled location, including those that do not vocalize during the survey. On the other hand, if the counts are conducted on successive days, for example, then we recognize that many individual birds may use the area subjected to our sampling, but that not all of them may be present at the exact time of a specific survey. In this sampling situation, abundance estimates produced by the *N*-mixture approach should include the superpopulation of birds that use the sampled location, even if some of these birds did not visit the location during the exact times of our repeat surveys.

The above 4 approaches to estimation of abundance are not the only ones developed for use with direct observations of animals. They were selected for discussion because they are widely used and because they illustrate the important point that all methods do not estimate the same "abundance." Instead, the abundance estimate may or may not include animals that are in the area exposed to sampling during the survey efforts, but that are not available for detection during this period. Similarly, the estimate may or may not include animals that use the area that is sampled, but do not use it at the exact times when we conduct our surveys. The central point is that the investigator must be aware of these differences in the quantities estimated by different methods and select an approach that is appropriate for the specific conservation program (see discussion in Nichols et al. 2009). Another key point is that various combinations of these approaches may permit the separate estimation of these various components of detection probability, in applications for which such decomposition would be useful (Farnsworth et al. 2005; Alldredge et al. 2006, 2007; Kissling and Garton 2006; Nichols et al. 2009; Riddle et al. 2010). Finally, we note the existence of other approaches to abundance estimation that are based on direct observations, including marked subpopulation, sighting probability models, and bounded counts (e.g., Williams et al. 2002).

Marked individuals. A long-standing approach to the study of animal populations involves the capture and marking of individual animals (e.g., Lincoln 1930). Capture–recapture approaches entail multiple sample periods, with animals captured at each session. Each captured animal is provided with an individual mark and is identified and recorded in subsequent recaptures. The data produced by such a study are the capture histories, a string of "0's" and "1's," indicating no capture or capture, respectively. Thus, a history of (0, 1) indicates an animal that was not captured at period 1, but was caught at period 2 of a two-period study. Although historic applications nearly always required that investigators apply marks to each captured animal, if animals are individually identifiable based on natural or acquired marks, then capture–recapture models can be based on remote photography, for example (e.g., O'Connell et al. 2010).

Capture–recapture models are usually classified as applying to "closed" and "open" populations. Closed population models are based on the assumption that the different capture occasions are sampling the same, unchanging population of animals. Thus, closed models assume that there are no deaths, births, or movements in or out of the population between sampling occasions. Based on this assumption, for each animal captured at least once in a study, a sampling occasion with no capture means an animal that was available for capture that was missed. Traditional closed models thus focus on modeling different sources of variation in capture probability (e.g., time, behavioral response, individual heterogeneity; see Otis et al. 1978; White et al. 1982; Chao and Huggins 2005a, b). Estimates of capture probabilities are then combined with numbers of captures to estimate abundance, either directly or indirectly (e.g., Eq. 8.2).

Open population models are developed for sampling situations in which populations may change via birth, death, and movement, between sampling occasions. Modeling is more complicated, because there is always uncertainty about whether the animal is present during sampling occasions prior to its initial capture or following its final capture. However, capture history "0's" occurring at sampling occasions between the first and last capture are known to represent simply non-capture. So inference about capture probability, and thus abundance, is again possible with such data, and an important by-product is inference about survival, recruitment, and movement processes (e.g., Williams et al. 2002). Just as different estimation approaches based on direct observations of unmarked animals lead to different abundance estimates, different estimators for open populations lead to abundance estimates that represent quantities ranging from the number of animals present in the sampled location at a specific sampling occasion (e.g., Jolly 1965), to the number of animals that use the sampled area during at least some period of the study (e.g., Schwarz and Arnason 1996; Williams et al. 2002).

One difficulty with capture–recapture modeling is heterogeneous capture probabilities, the situation in which some individuals in the focal population are more likely to be captured than others. One primary reason for such heterogeneity is the general location of animals with respect to the locations of capture devices. Efford (2004) developed an approach to deal with this problem using an assumption similar to that used in distance sampling. The probability of an individual being caught in any specific trap was hypothesized to be a function of the distance between the animal's center of activity and that trap. If the investigator records as additional information the location of each specific capture, in addition to the identity of the individual animal, then Efford (2004) showed how to use these data to estimate abundance and density (abundance per unit area). These spatially explicit capture–recapture models have proven very useful and are gaining increased popularity (Borchers and Efford 2008; Royle and Young 2008; Royle et al. 2009). Most work on these models has dealt with closed populations although open population models have just been developed as well (Gardner et al. 2010; Royle and Gardner 2010).

Capture–recapture models, both traditional and spatially explicit, are by far the most commonly used approaches to abundance estimation based on captures of animals. However, other approaches are sometimes used, including trapping webs (Anderson et al. 1983; Buckland et al. 2001), removal and catch–effort models (e.g., Gould and Pollock 1997; Williams et al. 2002), and change-in-ratio methods (Udevitz and Pollock 1991; Williams et al. 2002). In all of these cases, the basis for inference is expression (8.2). As is the case for observation-based inference methods, the selection of which capture-based approach to use for abundance estimation and the corresponding field survey design will be dictated by the larger conservation program that the estimates will serve and associated parameter needs (conservation models frequently require survival estimates and thus use of open population models) and logistical and related issues.

8.3 Conclusions and Recommendations

Abundance estimates are not always needed for programs of animal conservation. Decisions about whether or not to undertake a monitoring program that delivers abundance estimates should be based on the larger conservation program that those estimates are intended to serve. Specific roles and uses of abundance estimates should be clearly identified. For example, AM of an animal population or community requires estimates of focal state variables for the purposes of (1) making state-dependent decisions, (2) assessing the degree to which conservation progress is being made, and (3) learning, via comparison of abundance estimates with model-based predictions. Selection of abundance as a focal state variable (as opposed to species richness or some other metric) is based on the premise that it is an appropriate state variable with respect to these roles in conservation. When abundance is selected as a state variable for which estimates are needed, then a question still remains about what estimation method to select. Numerous reasonable methods have been developed, and it is important to select an approach that adequately deals with the two central conceptual issues underlying abundance estimation: geographic variation and detectability. Beyond this basic recommendation, the key is to select a specific estimation method based on the explicit needs of the conservation program and on the logistical constraints imposed by that program.

In summary, the estimation of abundance is not a stand-alone activity that is inherently useful to programs of animal conservation. Instead, it is best viewed as a component embedded within a larger program of conservation or management. The approach to abundance estimation, and the associated sampling design that it requires, should be inherited directly from the larger conservation program and the various logistical constraints and issues that it implies. This kind of close linkage between abundance estimation and the larger conservation program will provide the greatest likelihood that resulting estimates will be more than "low-information observations" and become maximally useful to the conservation process.

References

Alldredge MW, Pollock KH, Simons TR (2006) Estimating detection probabilities from multiple-observer point counts. Auk 123:1172–1182

Alldredge MW, Pollock KH, Simons TR, Collazo JA, Shriner SA (2007) Time of detection method for estimating abundance from point count surveys. Auk 124:653–664

Anderson DR, Burnham KP, White GC, Otis DL (1983) Density estimation of small-mammal populations using a trapping web and distance sampling methods. Ecology 64:674–680

Bellman R (1957) Dynamic programming. Princeton University Press, Princeton, NJ

Borchers DL, Buckland ST, Zucchini W (2002) Estimating animal abundance: closed populations. Springer, London

Borchers DL, Efford MG (2008) Spatially explicit maximum likelihood methods for capture–recapture studies. Biometrics 64:377–385

Buckland ST, Anderson DR, Burnham KP, Laake JL, Borchers DL, Thomas L (2001) Introduction to distance sampling. Oxford University Press, Oxford, UK

Buckland ST, Anderson DR, Burnham KP, Laake JL, Borchers DL, Thomas L (2004) Advanced distance sampling. Oxford University Press, Oxford, UK

Caughley G (1977) Analysis of vertebrate populations. Wiley, New York

Caughley G (1994) Directions in conservation biology. J Anim Ecol 63:215–244

Chao A, Huggins RM (2005a) Modern closed-population capture–recapture models. In: Amstrup SC, McDonald TL, Manly BFJ (eds) Handbook of capture–recapture analysis. Princeton University Press, Princeton, NJ, pp 58–87

Chao A, Huggins RM (2005b) Classical closed-population capture–recapture models. In: Amstrup SC, McDonald TL, Manly BFJ (eds) Handbook of capture–recapture analysis. Princeton University Press, Princeton, NJ, pp 22–35

Cochran WG (1953) Sampling techniques. Wiley, New York, NY
Cook RD, Jacobson JO (1979) A design for estimating visibility bias in aerial surveys. Biometrics 35:735–742
Crick HQP, Dudley C, Glue DE, Thomson DL (1997) UK birds are laying eggs earlier. Nature 388:526
Crick HQP (2004) The impact of climate change on birds. Ibis 146(s1):48–56
Efford M (2004) Density estimation in live-trapping studies. Oikos 106:598–610
Farnsworth GL, Pollock KH, Nichols JD, Simons TR, Hines JE, Sauer JR (2002) A removal model for estimating detection probabilities from point count surveys. Auk 119:414–425
Farnsworth GL, Nichols JD, Sauer JR, Fancy FG, Pollock KH, Shriner SA, Simons TR (2005) Statistical approaches to the analysis of point count data: a little extra information can go a long way. In: Ralph CJ, Rich TD (eds.) Bird conservation implementation and integration in the Americas: proceedings of the 3rd international partners in flight conference, vol 2. Gen. Tech. Rep. PSW-GTR-191. Albany, CA: Pacific Southwest Research Station, Forest service, U.S. Department Agriculture, pp 736–743
Gardner B, Reppucci J, Lucherini M, Royle JA (2010) Spatially explicit inference for open populations: estimating demographic parameters from camera-trap studies. Ecology 91:3376–3383
Gould WR, Pollock KH (1997) Catch-effort maximum likelihood estimation of important population parameters. Can J Fisher Aqat Sc 54:890–897
Johnson DH (2008) In defense of indices: the case of bird surveys. J Wildl Manag 72:857–868
Johnson FA, Breinenger DR, Duncan BW, Nichols JD, Runge MR, Williams BK (2011) A Markov decision process for managing habitat for Florida scrub-jays. J Fish Wildl Manag 2:234–246
Jolly GM (1965) Explicit estimates from capture–recapture data with both death and immigration-stochastic model. Biometrika 52:225–247
Karanth KU, Gopalaswamy AM, Kumar NS, Vaidyanathan S, Nichols JD, MacKenzie DI (2011) Monitoring carnivore populations at landscape scales: occupancy modeling of tigers from sign surveys. J Appl Ecol 48:1048–1056
Kissling ML, Garton EO (2006) Estimating detection probability and density from point-count surveys: a combination of distance and double-observer sampling. Auk 123:735–752
Lancia RA, Nichols JD, Pollock KH (1994) Estimating the number of animals in wildlife populations. In: Bookhout T (ed) Research and management techniques for wildlife and habitats. The Wildlife Society, Bethesda, Maryland, pp 215–253
Lancia RA, Kendall WL, Pollock KH, Nichols JD (2005) Estimating the number of animals in wildlife populations. In: Braun CE (ed) Research and management techniques for wildlife and habitats. The Wildlife Society, Bethesda, Maryland, pp 106–153
Lincoln FC (1930) Calculating waterfowl abundance on the basis of banding returns. Department of Agriculture Circular Number 342, U.S. Department of Agriculture, Washington, DC
Nichols JD, Hines JE, Sauer JR, Fallon FW, Fallon JE, Heglund PJ (2000) A double-observer approach for estimating detection probability and abundance from point counts. Auk 117:393–408
Nichols JD, Williams BK (2006) Monitoring for conservation. Trends Ecol Evol 21:668–673
Nichols JD, Thomas L, Conn PB (2009) Inferences about landbird abundance from count data: recent advances and future directions. In: Thomson DL, Cooch EG, Conroy MJ (eds) Modeling demographic processes in marked populations. Springer, New York, pp 201–235
Nichols JD, Williams BK (2013) Adaptive management. In: El-Shaarawi AH, Piegorsch WW (eds.) Encyclopedia of environmetrics. Wiley, Chichester, U.K. doi:10.1002/9780470057339.vnn037
O'Connell AF Jr, Nichols JD, Karanth KU (eds) (2010) Camera-traps in animal ecology: methods and analyses. Springer, Tokyo
Otis DL, Burnham KP, White GC, Anderson DR (1978) Statistical inference from capture data on closed animal populations. Wildl Monogr 62:1–135
Pinto GRM, Ferraz KMPM, Couto HTZ, Verdade LM (2006) Detectability of capybaras in forested habitats. Biota Neotropica 6(1). http://www.biotaneotropica.org.br/v6n1/pt/abstract?article+bn01906012006

Platt JR (1964) Strong inference. Science 146:347–353

Pollock KH, Nichols JD, Simons TR, Farnsworth GR, Bailey LL, Sauer JR (2002) Large scale wildlife monitoring studies: statistical methods for design and analysis. Environmetrics 13:1–15

Riddle JD, Stanislav SJ, Pollock KH, Moorman DE, Perkins FS (2010) Separating components of the detection process with combined methods, an example with Northern Bobwhite. J Wildl Manag 74:1319–1325

Royle JA, Nichols JD (2003) Estimating abundance from repeated presence absence data or point counts. Ecology 84:777–790

Royle JA (2004) *N*-mixture models for estimating population size from spatially replicated counts. Biometrics 60:108–115

Royle JA, Young KV (2008) A hierarchical model for spatial capture–recapture data. Ecology 89:2281–2289

Royle JA, Nichols JD, Karanth KU, Gopalaswamy AM (2009) A hierarchical model for estimating density in camera trap studies. J Appl Ecol 46:118–127

Royle JA, Gardner B (2010) Hierarchical spatial capture–recapture models for estimating density from trapping arrays. In: O'Connell AF, Nichols JD, Karanth KU (eds) Camera traps in animal ecology. Springer, Tokyo, pp 163–190

Schwarz CJ, Arnason AN (1996) A general methodology for the analysis of capture–recapture experiments in open populations. Biometrics 52:860–873

Seber GAF (1982) The estimation of animal abundance and related parameters. MacMillan, New York

Thompson SK (2002) Sampling. Wiley, New York

Udevitz MS, Pollock KH (1991) Change-in-ratio estimators for populations with more than two subclasses. Biometrics 47:1531–1546

White GC, Anderson DR, Burnham KP, Otis DL (1982) Capture–recapture and removal methods for sampling closed populations. Los Alamos National Laboratory Publication LA-8787-NERP, Los Alamos, NM

Williams BK (1997) Approaches to the management of waterfowl under uncertainty. Wildl Soc Bull 25:714–720

Williams BK, Nichols JD, Conroy MJ (2002) Analysis and management of animal populations. Academic Press, San Diego

Williams BK, Szaro RC, Shapiro CD (2007) Adaptive management: the U.S. department of the interior technical guide. U.S. Department of the Interior, Washington, DC

Yoccoz NG, Nichols JD, Boulinier T (2001) Monitoring of biological diversity in space and time. Trends Ecol Evol 16:446–453

Chapter 9
Wildlife Surveys in Agricultural Landscapes: Terrestrial Medium- to Large-Sized Mammals

Maria Carolina Lyra-Jorge, Carla Gheler-Costa, Carlos I. Piña, Luís Miguel Rosalino and Luciano M. Verdade

Abstract Human-induced changes have drastically modified pristine environments, and the replacement of natural ecosystems constrained the composition and structure of communities that they are made of, due to the modification of ecological processes. Survey and monitoring sampling schemes have been defined and mostly applied to natural environments, which are characterized by a spatial and structural heterogeneity. However, their application to agroforestry areas should take into consideration that these environments are spatially more homogeneous, but present a temporal heterogeneity linked with the production cycles. In this chapter, we present a description of the assumptions, weaknesses, and strengths of the main methods used in surveying and monitoring medium and large mammals. Moreover, we advise researchers to the need to take into consideration the particularities of agroforestry landscapes and adapt

M. C. Lyra-Jorge (✉)
Curso de Ciências Biológicas, Universidade de Santo Amaro,
Rua Prof. Eneas de Siqueira Neto, 340, São Paulo, SP, 04829-300, Brazil
e-mail: carolina_ecologia@hotmail.com

C. Gheler-Costa
Universidade Sagrado Coração, Rua Arminda, 10-50, Bauru, SP 17011-160, Brazil
e-mail: cgheler@gmail.com

C. I. Piña
Centro de Investigaciones Científicas y Transferencia de Tecnología a la Producción, (CICYTTP-CONICET)/FCyT-UAdER/FCAL-UNER, Dr. Materi y España, CP 3105, Diamante, Entre Ríos, Argentina
e-mail: cidcarlos@infoaire.com.ar

L. M. Rosalino
Centro de Biologia Ambiental—Faculdade de Ciências de Lisboa, Universidade de Lisboa, Ed. 2, Campo Grande, 1749-016 Lisboa, Portugal
e-mail: lmrosalino@fc.ul.pt

L. M. Rosalino
Laboratório de Ecologia Isotópica—CENA/USP, C. P. 96, Piracicaba, SP 13416-000, Brazil

L. M. Verdade
Centro de Energia Nuclear na Agricultura, Universidade de São Paulo, Caixa Postal 96, Piracicaba, SP 13416-000, Brazil
e-mail: lmverdade@usp.br

L. M. Verdade et al. (eds.), *Applied Ecology and Human Dimensions in Biological Conservation*, DOI: 10.1007/978-3-642-54751-5_9,

the mentioned methods to assure the representativeness of the collected data and the accuracy of the detected patterns.

9.1 A New Landscape

Over the last decades, land use intensification has induced important changes in the terrestrial ecosystems throughout the world, such as the destruction of natural habitats, the fragmentation and isolation of native patches, and the introduction of exotic species, some of which became invasive (Turner and Meyer 1994). In Europe, particularly Switzerland, almost all wetlands have been converted into anthropic landscapes in the last 150 years. But this pattern is not exclusive of areas where humans are present for centuries or millenia. In younger countries, such as Australia, the expansion of pasture lands and sheep grazing were responsible for a 10 % reduction in the natural land cover in some regions (Henle et al. 2004). In Brazil, only 12 % of the Atlantic rainforest and 20 % of the Cerrado present when the first Portuguese sailors reached this region still subsist (Ribeiro et al. 2009). This landscape conversion may have huge negative impacts on native flora and fauna, and therefore, habitat destruction and fragmentation are considered two of the major causes of the increased species extinction rates in the last decsades (Daily et al. 2003).

Fragmented landscapes can be important in biodiversity conservation if they still maintain their functional connectivity, i.e., a link between fragmented habitats, due to their relative spatial proximity or due to the landscape matrix permeability to species movements (With et al. 1997). Thus, the matrix quality, especially those composed of agroforestry systems, is crucial to promote connectivity between patches of native vegetation (With et al. 1997). When the matrix is composed of agriculture lands, it often does not act as a non-habitat structure for native species, since some species manage to take advantage of the resources it provides and uses regularly (Gheler-Costa et al. 2012). In such situations, these environments still maintain an intrinsic value in the conservation process. Several studies have showed that coffee plantations in Mexico (e.g., Moguel and Toledo 1999), banana and coconut plantations in Costa Rica (e.g., Harvey et al. 2006), cocoa plantations in Brazil (e.g., Faria et al. 2006), subsistence agriculture in Nepal (e.g., Acharya 2006), and silvicultural areas of Mediterranean Europe (e.g., Rosalino et al. 2005) and Brazil (e.g., Lyra-Jorge et al. 2008a; Gheler-Costa et al. 2012; Martin et al. 2012) are regularly used by the native fauna in their ecological processes framework (Fig. 9.1).

Habitat quality, in terms of quality and quantity of resources it supports, determines the persistence and abundance of flora and fauna species in particular regions, whatever the scale considered (Fahrig and Merriam 1995). However, recent studies have showed that many animal populations have the ability to acclimate or adapt to the changes in the original habitats (Morán-López et al. 2006; McDougall et al. 2006, Sánchez-Hernández et al. 2001, Tabeni and Ojeda 2005, Rosalino et al 2014). Several species have even managed to adapt to urban areas, changing their ecological and behavioral patterns to survive in such anthropic environments (e.g., increase in the population density together

Fig. 9.1 *Mazama guazoubira* female and cub in a Brazilian *Eucalyptus* plantation

Fig. 9.2 Cork oak (*Quercus suber*) forest in southern Portugal

with a decrease in individual territory sizes, reduction in the migratory behavior, extension of the reproductive period, alteration of the activity rhythms and diet, and an increase in species tolerance toward man Luniak 2004).

Landscape fragmentation, associated with the establishment of agroforestry systems, together with the consequent changes in the ecological processes and species local extinctions, enhances the need to integrate human's requirements into the preservation of essential ecosystem processes. The history of agriculture in Europe shows the development of several regional agro-pastoral systems, characterized by a strong interaction between nature and human cultures. These examples can become the baseline for the implementation of a sustainable agriculture or for the convergence of aggressive production schemes into nature-friendly but competitive systems (Bignal 1998). A good example of such a scheme is the oak forests of southern Iberian Peninsula—called "montado" in Portugal (Fig. 9.2) and "alcornocales" in Spain (Grove and Rackham 2003). These are one of the last agro-silvo-pastoral systems in Europe, characterized by diverse and complementary productions (agriculture, cattle breeding, and forestry) associated with a high biodiversity (Diáz et al. 1997), globally managed in a sustainable manner.

Such examples have proved that the preservation of biodiversity can be achieved in agriculture systems if those systems are able to incorporate conservation concepts and if conservationist can consider agriculture systems as areas that can be used by wildlife (Vandermeer and Perfecto 1997, Verdade et al. 2014a). But to assess the role of agroforestry landscapes for conservation, we need to adapt the survey and monitoring methodologies that have been developed in pristine areas to the particularities of anthropic areas.

9.2 Wildlife Survey Methods

Population and community monitoring processes allow the assessment of species richness, species distribution, abundance and density, and habitat use, among other parameters that vary according to the researchers' needs and methods used (Fig. 9.3).

Due to the species habits (e.g., nocturnal) and elusive characters, which make direct observations or even successful captures difficult, studies focused on medium- or large-sized mammals often use indirect methods based on signs of presence, such as feces, footprints, prey remains, etc. (Sutherland 2006). However, although being easy to implement, these methods' accuracy may be biased, since their successful use depends on the observer's experience (Smallwood and Fitzhugh 1995). Among these indirect methods, those based on unstandardized footprint detection are probably the oldest used to study medium/large mammals (Bider 1968). Later, and in combination with footprints, researchers started to use other signs of presence to assess species occurrence, which included feces, nests/burrows, runways, or claw mark detection. More recently, standardized methods have been developed (often using the above-mentioned signs of presence), namely line transects (Rudran et al. 1996), counting calls (Fuller and Sampson 1988), camera trapping (Wemmer et al. 1996), track plots and scent stations (Zielinski and Kucera 1995), ethnobiology methods (Comin et al. 2009), hair tubes or hair catchers, and molecular approaches (Zielinski and Kucera 1995; Sutherland 2006; Long et al. 2008).

Whatever the selected method, the levels of accuracy and precision will vary between techniques, species, and environments (Verdade et al. 2012), as well as their cost–benefit relations (Gaidet-Drapier et al. 2006). Therefore, choosing the more appropriate method to survey medium–large-sized mammals should be guided by the study objectives (i.e., what data are needed), species involved, landscape characteristics, experience of the researcher, and logistic and financial support available. Moreover, caution should be used in data analysis, as the results obtained depend on the methods' detectability, accuracy, and observer's experience. A non-detection of a particular species does not imply that it is absent

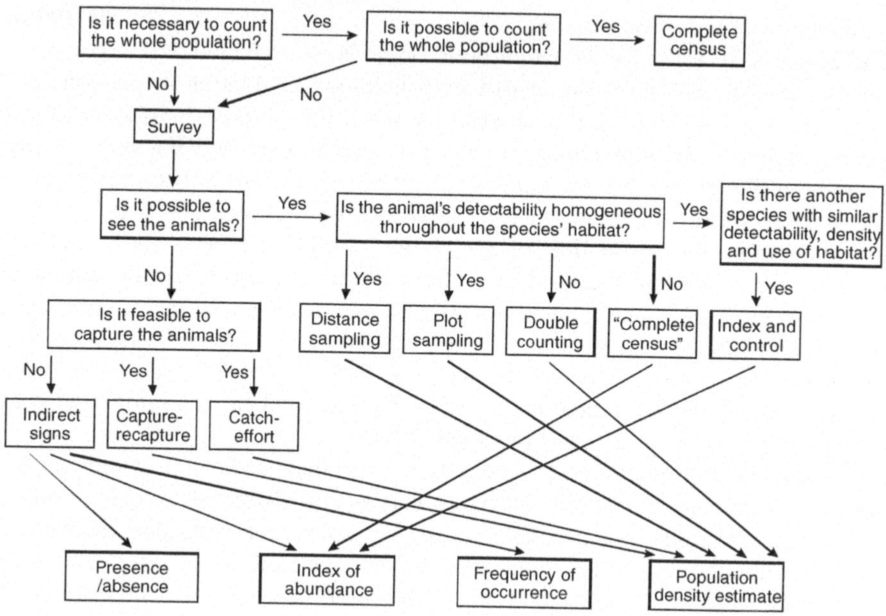

Fig. 9.3 Field methods and the questions they can answer [adapted from Verdade et al. (2012) and Lancia et al. (1996)]

from the study region. It could be highly difficult to detect (e.g., cryptic species) or the study sample unit location does not overlap with the habitats preferentially used by the species. Even if the selected method is highly effective in detecting the majority of the species inhabiting the area (which could be easily tested by an incidence-based species accumulation curves Soberón and Llorente 1993), the difference in abundance between them can also be biased by the methods detectability (e.g., different defecation behaviors—latrines vs. non-latrines—will influence number of feces detected in road transects and consequently abundance indexes based on feces counts). Finally, it is important to refer that species' abundance in nature is usually quite different, with a common pattern: common being rare and rare being common (Verdade et al. 2014b).

Studies focused on wildlife monitoring are often limited by budgets and therefore, it is essential to assess the methods' performance, it costs, and cost–benefit relations. For example, the use of genetic tools can provide accurate data, although the regular and widespread use of this approach is limited by the associated high financial costs (Long et al. 2008). Camera trapping and line footprint surveys (e.g., line transects or track plot) are nowadays two of the most used methods in wildlife monitoring. This high use derives from its easy implementation and data collection in the field. While footprint's survey depends mainly on the researcher's experience and on weather and soil condition, camera trapping is far less affected by those factors, which implies a lower maintenance effort (e.g., one camera can be active in the field for several weeks without maintenance), and produces higher-accuracy results (Wemmer et al. 1996). Lyra-Jorge et al. (2008b) compared both methods and concluded that considering the benefits and shortcomings of both methods in relation to performance and costs, track plot method is satisfactory when the purpose is to assess local species richness through quick surveys and under a limited budget. Camera trapping, although being more expensive and biased toward larger animals, permits precise species identification, the investigation of species activity patterns, and sometimes the estimation of population density (through individualized records with or without the need for individual recognition) (Voss and Emmons 1996; Srbeck-Araújo and Chiarello 2007; Rowcliffe et al. 2008). It is an efficient method especially to detect cryptic animals which are solitary, living in low density or in small groups (Carbone et al. 2001). Both methods can be used together, enhancing field data quality (e.g., increasing data resolution) and providing complementary results (Long et al. 2008).

The data quality of all the ecological parameters that can be collected by applying the several survey methods available may also be affected by extrinsic factors such as the study area accessibility and the interaction with local inhabitants (Gaidet-Drapier et al. 2006). For example, in populated areas, interviews with local residents should be a method more often used in wildlife inventory projects (Huntington 2000). Although the interviewer skill in identifying misleading information together with the willingness of residents to provide such information may bias the collected data, local inhabitants may supply, in a rapid and without additional cost process, a list of species present in the study area. This list should be compared with the data provided by the other methodologies used in the project, to access its accuracy and complementarity.

9.3 Limiting Factors of Wildlife Surveys in Agriculture Landscapes

Wildlife survey approaches in pristine or natural environments should differ from those implemented in agroforestry landscapes. Natural landscapes present a higher spatial heterogeneity, since they are composed of diverse habitats with different vegetation structures (different vegetation strata) and, therefore, present higher β diversity. Inversely, agroforestry landscapes are often monocultures (e.g., sugarcane or *Eucalyptus* plantations), with a more homogeneous spatial structure (Fig. 9.4).

The structure of agroforestry landscapes is highly dependent on the production cycles; therefore, it varies seasonally, from plantation to harvesting. Thus, the environment in such landscapes presents a high temporal variation. For example, in *Eucalyptus* plantations, the first stands' phase presents a typical shrub-like structure, which evolves toward a forest system in 6–7 years in the Neotropics or 9–10 in temperate regions, before harvesting, where these forests are reduced to bare soil areas. In such situations, the wildlife survey sampling design should take into consideration such variation. In a pristine area, the heterogeneity is mainly spatially determined.

Fig. 9.4 Sugarcane plantation

Therefore, whatever the method selected, the researcher could assess species richness and distribution by sampling the area in unique (or few) sampling events, provided that sampling plots (preferentially with a standardized spatial distribution) covering all or at least the most abundant land covers. Such design allows the detection of habitat generalist as well as habitat specialist species. However, as agricultural crops have a high temporal heterogeneity, sampling design should be planned in order to detect such short term variations which are due to ecological not sampling processes (see Preston 1960). Moreover, results analysis and discussion should always have in mind that the detected ecological processes and patterns are not only determined by present-day conditions, but also mostly by the history of human presence and activities in the region (Lunt and Spooner 2005, Balée 2014) and by the acclimation and adaptation strategies adopted by the species to cope with those man-induced changes (Rosalino et al. 2014).

For these reasons, species monitoring in agroforestry landscapes should incorporate a temporal scale (i.e., several sampling events along the production cycle), so results can reflect the community evolution and the influence of the production cycle upon the detected patterns. However, a standardization of the sampling process should be maintained to assure the robustness of the seasonal comparisons. Often, in agroforestry areas, researchers or research groups have implemented short-term studies (often associated with the need to comply with academic deadlines—e.g., dissertation or thesis). However, due to the particular temporal variation of such systems, the representativeness of the collected data may be questionable and the detected patterns misleading. Thus, assuming that the same sampling method and design can equally sample pristine and agroforestry landscapes is not correct.

Every method developed to survey medium/large mammalian species has strengths and weaknesses (Table 9.1). The selection of the most appropriate method for altered landscapes should be guided by each study's specific characteristics, considering every specific bias associated with the techniques. However, understanding the meaning of the collected data in the context of the landscape history is one of the greatest challenges a research can face.

Table 9.1 Assumptions, stregths and weaknesses of medium/large mammals survey methods

Methods	Assumptions	Weaknesses	Strengths	Examples
Interviews / questionnaires	• Presence of human communities	• Species identification errors associated with the observer • Omission of information (e.g., game) • Monitored area limited to that used by humans	• Easy to implement • Cheap • Allow monitoring a huge number of species • Rapid results • Does not need specialized equipment	Yamada et al. (2003) Borralho et al. (1995)
Track plots	• All species produce recognizable footprints • Correlation between footprints' abundance and population size	• Depends on the substrate • It is influenced by vehicle and human movements • Depends on the researcher experience • It is influenced by climate conditions (e.g., rain) • Low accuracy	• Easy to implement • Cheap • It is possible to use surrounding land • Rapid results • Does not need specialized equipment	Lyra-Jorge et al. (2008b) Ascensão and Mira (2007) Zoellick et al. (2005)
Camera trap	• All animals moving across the camera activation area will be photographed	• High cost • The equipment can be stolen • Detection is limited by the distance to the camera (focal distance) • It is influenced by vehicle and human movements • Detection depends on the animals' size • Some species learn to avoid the camera	• Does not depend on the researcher's experience • Does not imply the permanence of the researcher in the field during all monitoring period • Allows the registration of the animals' behaviors and activity rhythms • Allows saving activity and behavioral videos • High accurracy in species identification	Sarmento et al. (2010) Manzo et al. (2011) Lyra-Jorge et al. (2008a)

(continued)

Table 9.1 (continued)

Methods	Assumptions	Weaknesses	Strengths	Examples
Feces collection	• Feces of all species present in the study area can be found and identified	• Depends on the researcher's experience • It is influenced by climate conditions (e.g., rain) • Low accuracy	• Easy to implement • Cheap • Does not need specialized equipment • Rapid results	Sales-Luís et al. (2012) Rosalino et al. (2009)
Linear transects	• Environments allow for a large field of vision • All mammals have the same probability of being located • There is a correlation between animal counts and population size	• Observations are biased by closed environments • Observer saturation in face of a high number of animals • Time consuming • Non-random sampling related to roads or other paths	• Cheap • Does not need specialized equipment	Cuesta et al. (2003) Cuarón et al. (2004) Ruette et al. (2003) Hanby and Bygott (1979)
Counting calls	• The species vocalizations are common and identifiable	• Mainly used with primates, which are usually not found in non-forest environments • Depends on the researcher's experience	• Accurate identification • Allows to determine the number of animals present • Allows the confirmation of reproduction (e.g. Wolf)	Price (1994) Fuller and Sampson (1988) Waser (1977)
Capture–recapture	• All animals have the same probability of being captured	• Implies the capture and handling of the animals • High fieldwork effort • Implies the presence of a veterinary • Needs specialized equipment	• High accuracy in animal and species identification • Allows the collection of diversified data (e.g. parasites, reproduction, age, etc.)	Hawkins and Racey (2005) Coonan et al. (2005) Hansen et al. (2004)

(continued)

Table 9.1 (continued)

Methods	Assumptions	Weaknesses	Strengths	Examples
Scent stations	• Scent is one of the species main senses	• Lures (scents) can be species specific (i.e., some species are more attracted than others) • It is influenced by climate conditions (e.g., rain) • Needs a frequent renovation of lures • Repulsive effect of bigger/predators scent marks upon smaller/prey species	• Easy to implement • Cheap • Can be species specific, allowing the optimization of species detection (specific lures)	Travaini et al. (1996) Conner et al. (1983) Linhart and Knowlton (1975)
Radiotracking	• The study area permits a continuous animals' monitoring with radio signals and/or full satellite cover	• High cost • Implies the capture and handling of the animals • Implies the presence of a veterinary • High field work effort • The use of GPS transmitters are limited by the animals weight	• Allows continuous monitoring • High accuracy in animal's location • Allows the simultaneous monitoring of several animals • High amount of collectable data	Soisalo and Cavalcanti (2006) Rosalino et al. (2004) Zielinski et al. (2004)
Genetic markers		• High cost • High contamination risk of the samples • Needs specialized equipment • Researchers need to have a lab technical specialization • Low analytical success for some biological samples (e.g. old faeces) • Implies the availability (or development) or specific genetic markers	• Allows individual identification • High accuracy • Allows the simultaneous monitoring of several species/individuals • High amount of collectable data	Long et al. (2011) Palomares et al. (2002) Prigioni et al. (2006)

9.4 Final Remarks

The data presented in the previous section regarding the assumptions and limitations of monitoring medium–large size mammals on agro-forestry landscapes may have raised more concerns than pointed out solutions. Facing these difficulties as new challenges it is fundamental that researchers focus their studies on identifying the ecological adaptations of mammals to these new and permanent changing landscapes. Thus, we suggest the following:

1. Mammalian research groups should coordinate their efforts to implement long-term studies, encompassing standardized and systematic data collection protocols, whose results should be comparable, to provide managers with answers to the decision making process.
2. Sampling design should include a multiple approach by considering several sampling methods, so the obtained results could be complementary and the final output more accurate.
3. The analysis of how the mammalian community evolves over time in agroforestry landscapes should be a priority, since although these areas are often considered poor in biodiversity they can still support some mammal species.
4. Population biology and fitness studies should also be implemented, especially those that can provide density estimations and a fitness assessment of animals living in agroforestry systems. This data will help researchers to assess the real conservation role of such environments to mammals.
5. Whenever possible, researchers should include genetic tools in their methodological approach due to its high accuracy.
6. The implementation of studies that analyze the changes and ecological adaptation of species to agroforestry systems should also be considered.

Mammals, as most of the vertebrates, face in many regions of the world multiple threats, often associated with habitat fragmentation whose effects might be enhanced by global climate changes. These effects have led many species to reduce drastically their distribution area, thus being urgent to define conservation strategies that may allow their survival (Lindenmayer and Burgman 2005). However, while some species have managed to recover due to human rural emigration and landscape natural revegetation, especially in Europe, others managed to acclimate or adapt to anthropic environments, managing to survive and reproduce in habitats considered suboptimal (Verdade et al. 2011). The coexistence of such different patterns, often sympatric, should motivate researchers to produce relevant, robust, and systematic information on how these species use the landscape, to allow the identification of the processes that support those patterns. For this to happen it is crucial that the sampling methods selected by researchers take into consideration not only the species characteristics, but also the study objectives, the landscape features and the logistical feasibility of the methodology. We hope that the present chapter might help researchers in this task.

References

Acharya KP (2006) Linking tree on farms with biodiversity conservation in subsistence farming in Nepal. Biodiver Conserv 15:631–646

Ascensão F, Mira A (2007) Factors affecting culvert use by vertebrates along two stretches of road in southern Portugal. Ecol Res 22:57–66

Balée W (2014) Historical ecology and the explanation of diversity: Amazonian Case Studies. In: Verdade LM, Lyra-Jorge MC, Piña CI (eds) Applied ecology and human dimensions in biological conservation. Springer, Heidelberg

Bider JR (1968) Animal activity in uncontrolled terrestrial communities as determined by sand transect technique. Ecol Monogr 38:269–308

Bignal EM (1998) Using an ecological understanding of farmland to reconcile nature conservation requirements, EU agriculture policy and world trade agreements. J Appl Ecol 35:949–954

Borralho R, Rego F, Palomares F, Hora A (1995) The distribution of the Egyptian mongoose *Herpestes ichneumon* (L.) in Portugal. Mamm Rev 25:229–236

Carbone C, Christie S, Conforti K, Coulson T, Franklin N, Ginsberg JRI (2001) The use of photographic rates to estimate densities of tigers and other cryptic mammals. Anim Conserv 4:75–79

Conner MC, Labisky RF, Progulske DR (1983) Scent-station indices as measures of population abundance for bobcats, raccoons, gray foxes, and opossums. Wildl Soc Bull 11:146–152

Comin FH, Gheler-Costa C, Verdade LM, Garavello MEP (2009) Relações e conflitos na conservação ambiental da bacia do rio Passa-Cinco, São Paulo, Brasil. OLAM 9:254–274

Coonan TJ, Schwemm CA, Roemer GW, Garcelon DK, Munson L (2005) Decline of an Island fox subspecies to near extinction. SW Nat 50:32–41

Cuarón AD, Martínez-Morales MA, McFadden KW, Valenzuela D, Gompper ME (2004) The status of dwarf carnivores on Cozumel Island, Mexico. Biodiver Conserv 13:317–331

Cuesta F, Peralvo MF, Frank T, van Manen FT (2003) Andean bear habitat use in the Oyacachi river Basin, Ecuador. Ursus 14:198–209

Diáz M, Campos P, Pulido FJ (1997) The Spanish dehesas: a diversity in land-use and wildlife. In: Pain DJ, Pienkowski MW (eds) Farming and birds in Europe: the common agriculture policy and its implication for birds conservation. Academic Press LDA, London, pp 178–209

Daily G, Ceballos G, Pacheco J, Suzán G, Sanchéz-Azofeifa A (2003) Countryside biogeography of Neotropical mammals: conservation opportunities in agricultural landscapes of Costa Rica. Conserv Biol 17:1814–1826

Faria D, Laps RR, Baumgarten J, Cetra M (2006) Bat and bird assemblages from forests and shade cacao plantations in two contrasting landscapes in the Atlantic forest of southern Bahia, Brazil. Biodiver Conserv 15:587–612

Fahrig L, Merriam G (1995) Conservation of fragmentation populations. In: Ehrenfeld D (ed) Readings from conservation biology—the landscape perspective. Blackwell, Cambridge, pp 16–25

Fuller TK, Sampson BA (1988) Evaluation of a simulated howling survey for wolves. J Wildl Manag 52:60–63

Gaidet-Drapier N, Fritz H, Bougarel M, Renaud PC, Poilecot P, Chardonet P, Coid C, Poulet D, Le Bel S (2006) Cost and efficiency of large mammal census techniques: comparison of methods for a participatory approach in a communal area, Zimbabwe. Biodiver Conserv 15:735–754

Gheler-Costa C, Vettorazzi CA, Pardini R, Verdade LM (2012) The distribution and abundance of small mammals in agroecosystems of south-eastern Brazil. Mammalia 76:185–191

Grove AT, Rackham O (2003) The nature of Mediterranean Europe. An ecological history. Yale University Press, New Haven

Hanby JP, Bygott D (1979) Population changes in lions and other predators. In: Sinclair AR, Norton-Griffiths M (eds) Serengeti: dynamics of an ecosystem. University of Chicago Press, Chicago, pp 249–262

Hansen LA, Mathews NE, Lee BAV, Lutz R (2004) Population characteristics, survival rates, and causes of mortality of striped skunks (*Mephitis mephitis*) on the southern High Plains, Texas. SW Nat 49:54–60

Harvey C, Gonzalez J, Somarriba E (2006) Dung beetle and terrestrial mammal diversity in forests, indigenous agroforestry systems and plantain monocultures in Talamanca, Costa Rica. Biodiver Conserv 15:555–585

Hawkins CE, Racey PA (2005) Low population density of a tropical forest carnivore, *Cryptoprocta ferox*: implications for protected area management. Oryx 39:35–43

Henle K, Linddenmayer DB, Margules CR, Saunders DA, Wissel C (2004) Species survival in fragmented landscapes: where are we now? Biodiver Conserv 13:1–8

Huntington HP (2000) Using traditional ecological knowledge in science: methods and applications. Ecol Appl 10:1270–1274

Lancia RA, Nichols JD, Pollock KH (1996) Estimating the number of animals in wildlife populations. In: Bookhout TA (ed) Research and management techniques for wildlife and habitats, 5th edn. The Wildlife Society, Bethesda, pp 215–253

Lindenmayer D, Burgman M (2005) Practical conservation biology. CSIRO Publishing, Collingwood

Linhart SB, Knowlton FF (1975) Determining the relative abundance of coyotes by scent station lines. Wildl Soc Bull 3(3):119–124

Long RA, MacKay P, Zielinski WJ, Ray JC (2008) Noninvasive survey methods for carnivores. Island Press, Washington, DC

Long R, Donovan T, MacKay P, Zielinski W, Buzas J (2011) Predicting carnivore occurrence with non-invasive surveys and occupancy modelling. Landsc Ecol 26:327–340

Luniak M (2004) Synurbization—adaptation of animal wildlife to urban development. In: Proceedings of the 4th international symposium on urban wildlife conservation, Tucson, Arizona, University of Tucson, Tucson 1–5 May 1999, pp 50–55

Lunt ID, Spooner PG (2005) Using historical ecology to understand patterns of biodiversity in fragmented agricultural landscapes. J Biogeogr 32:1859–1873

Lyra-Jorge MC, Ciocheti G, Pivello VR (2008a) Carnivore mammals in a fragmented landscape in northeast of São Paulo State, Brazil. Biodiver Conserv 17:1573–1580

Lyra-Jorge MC, Ciocheti G, Pivello VR, Meirelles ST (2008b) Comparing methods for sampling large—and medium-sized mammals: camera traps and track plots. Eur J Wildl Res 54:739–744

Manzo E, Bartolommei P, Rowcliffe JM, Cozzolino R (2011) Estimation of population density of European pine marten in central Italy using camera trapping. Acta Theriol 57:165–172

Martin PS, Gheler-Costa C, Lopes PC, Rosalino LM, Verdade LM (2012) Terrestrial non-volant small mammals in agro-silvicultural landscapes of Southeastern Brazil. For Ecol Manag 282:185–195

McDougall PT, Réale D, Sol D, Reader SM (2006) Wildlife conservation and animal temperament: causes and consequences of evolutionary change for captive reintroduced and wild populations. Anim Conserv 9:39–48

Moguel P, Toledo VM (1999) Biodiversity conservation in traditional coffee systems of Mexico. Conserv Biol 13:11–21

Morán-López R, Guzmán JM, Borrego EC, Sánchez AV (2006) Nest-site selection of endangered cinereous vulture populations affect by anthropogenic disturbance: present and future conservation implications. Anim Conserv 9:29–37

Palomares F, Godoy JA, Piriz A, O'Brien SJ, Johnson WE (2002) Faecal genetic analysis to determine the presence and distribution of elusive carnivores: design and feasibility for the Iberian lynx. Mol Ecol 11:2171–2182

Preston FW (1960) Time and space and the variation of species. Ecology 41:612–627

Price K (1994) Center-edge effect in red squirrels: evidence from playback experiments. J Mammal 75:545–548

Prigioni C, Remonti L, Balestrieri A, Sgrosso S, Priore G, Mucci N, Randi EI (2006) Estimation of European otter (*Lutra lutra*) population size by fecal DNA typing in southern Italy. J Mammal 87:855–858

Ribeiro MC, Metzger JP, Martensen AC, Ponzon FJ, Hirota MM (2009) The Brazilian Atlantic Forest: how much is left, and how is the remaining forest distributed? Implications for conservation. Biolog Conserv 142:1141–1153
Rosalino LM, Macdonald DW, Santos-Reis M (2004) Spatial structure and land-cover use in a lowdensity Mediterranean population of Eurasian badgers. Can J Zool 82:1493–1502
Rosalino LM, Lyra-Jorge MC, Verdade LM (2014) Adaptation and evolution in changing environments. In: Verdade LM, Lyra-Jorge MC, Piña, CI (eds) Applied ecology and human dimensions in biological conservation, Springer-Verlag, Berlin
Rosalino LM, Macdonald DW, Santos-Reis M (2005) Resource dispersion and badger population density in Mediterranean woodlands: is food, water or geology the limiting factor? Oikos 110:441–452
Rosalino LM, Rosário J, Santos-Reis M (2009) The role of habitat patches on mammalian diversity in cork oak agroforestry systems. Acta Oecol 35:507–512
Ruette S, Stahl P, Albaret M (2003) Applying distance-sampling methods to spotlight counts of red foxes. J Appl Ecol 40:32–43
Rowcliffe JM, Field J, Turvey ST, Carbone C (2008) Estimating animal density using camera traps without the need for individual recognition. J Appl Ecol 45:1228–1236
Rudran R, Kunz TH, Southwell C, Jarman P, Smith A (1996) Observational techniques for nonvolant mammals. In: Wilson DE, Cole FR, Nichols JD, Rudran R (eds) Foster MS measuring and monitoring biological diversity. standard methods for mammals. Smithsonian Institution Press, Washington, DC, pp 81–103
Sales-Luís T, Bissonette JA, Santos-Reis M (2012) Conservation of Mediterranean otters: the influence of map scale resolution. Biodiver Conserv 21:2061–2073
Sánchez-Hernández C, Romero-Almaraz ML, Colín-Martinez H, García-Estrada C (2001) Mamíferos de cuatro áreas com diferente grado de alteración en el sureste de México. Acta Zool Mex 84:35–48
Sarmento PB, Cruz JP, Eira CI, Fonseca C (2010) Habitat selection and abundance of common genets *Genetta genetta* using camera capture-mark recapture data. Eur J Wildl Res 56:59–66
Smallwood KS, Fitzhugh EL (1995) A track count for estimating mountain lion *Felis concolor californica* population trend. Biolog Conserv 71:251–259
Soisalo MK, Cavalcanti SM (2006) Estimating the density of a jaguar population in the Brazilian Pantanal using camera-traps and capture–recapture sampling in combination with GPS radiotelemetry. Biolog Conserv 129:487–496
Soberón J, Llorente J (1993) The use of species accumulation functions for the prediction of species richness. Conserv Biol 7:480–488
Srbek-Araújo AC, Chiarello AG (2007) Armadilhas fotográficas na amostragem de mamíferos: considerações metodológicas e comparação de equipamentos. Rev Bras Zool 24:647–656
Sutherland WJ (2006) Ecological census techniques—a handbook. Cambridge University Press, Cambridge
Tabeni S, Ojeda RA (2005) Ecology of the desert small mammals in disturbed and undisturbed habitats. J Mammal 70:416–420
Travaini A, Laffitte R, Delibes M (1996) Determining the relative abundance of European red foxes by scent-station methodology. Wildl Soc Bull 24:500–504
Turner BL II, Meyer WB (1994) Global land use and land cover change: an overview. In: Meyer WB, Turner BL II (eds) Changes in land use and land cover: a global perspective. Cambridge University Press, Cambridge, pp 3–12
Vandermeer J, Perfecto I (1997) The agroecosystem: a need for the conservation biologist's lens. Conserv Biol 11:591–592
Verdade LM, Moreira JR, Ferraz KMPMB (2012) Counting capybaras. In: Moreira JR, Ferraz KMPMB, Herrera EA, Macdonald DW (eds) Capybara: biology, use and conservation of an exceptional Neotropical species. Springer, New York, pp 357–370
Verdade LM, Rosalino LM, Gheler-Costa C, Pedroso NM, Lyra-Jorge MC (2011) Adaptation of mesocarnivores (Mammalia: Carnivora) to agricultural landscapes of Mediterranean Europe and southeastern Brazil: a trophic perspective. In: Rosalino LM, Gheler-Costa C (eds) Middle-sized carnivores in agricultural landscapes. Nova Science Publishers, New York, pp 1–38

Verdade LM, Penteado M, Gheler-Costa C, Dotta G, Rosalino LM, Pivello VR, Lyra-Jorge MC (2014a) The conservation value of agricultural landscapes. In: Verdade LM, Lyra-Jorge MC, Piña CI (eds) Applied ecology and human dimensions in biological conservation. Springer, Heidelberg

Verdade LM, Lyra-Jorge MC, Piña CI (2014b) Redirections in conservation biology. In: Verdade LM, Lyra-Jorge MC, Piña CI (eds) Applied ecology and human dimensions in biological conservation. Springer, Heidelberg

Voss RS, Emmons LH (1996) Mammalian diversity in Neotropical lowland rainforest: a preliminary assessment. Bull Am Mus Nat Hist 230:1–115

Waser PM (1977) Individual recognition, intragroup cohesion and intergroup spacing: evidence from sound playback to forest monkeys. Behaviour 60:28–74

Wemmer C, Kunz TH, Lundie-Jenkins G, McShea WJ (1996) Mammalian sings. In: Wilson DE, Cole FR, Nichols JD, Rudran R, Foster MS (eds) Measuring and monitoring biological diversity. Smithsonian Institution Press, Washington, Standard methods for mammals, pp 157–176

With K, Gardner R, Turner M (1997) Landscape connectivity and population distributions in heterogeneous environments. Oikos 78:151–169

Yamada K, Elith J, McCarthy M, Zerger A (2003) Eliciting and integrating expert knowledge for wildlife habitat modelling. Ecol Model 165:251–264

Zielinski WJ, Kucera TE (1995) American marten, fisher, lynx, and wolverine: survey methods for their detection. General technical report PSW GTR-157. United States Department of Agriculture, Forest Service, Berkeley

Zielinski WJ, Truex RL, Schmidt GA, Schlexer FV, Schmidt KN, Barrett RH (2004) Home range characteristics of fishers in California. J Mammal 85:649–657

Zoellick BW, Ulmschnieder HM, Stanley AW (2005) Distribution and composition of mammalian predators along the Snake river in southwestern Idaho. NW Sci 79:265–272

Chapter 10
Point Counts Method for Bird Surveys in Agroecosystems of the State of São Paulo, Southeastern Brazil

Marli Penteado, Wesley R. Silva and Luciano M. Verdade

Abstract The point counts method has been developed for, and extensively used in, forest habitats for bird surveys. Although the method has already been applied to anthropic habitats, its efficacy has never been tested in such circumstance. The main goal of this study was to test this method in different agricultural habitats. We surveyed birds in 16 study sites of the following types of habitat of Passa-Cinco river basin (between latitudes 22°05′ and 22°30′S, and longitudes 47°30′ and 47°50′W) in the state of São Paulo, Southeastern Brazil, from September 2003 to January 2005: native forest fragments, *Eucalyptus* and sugarcane plantations, and exotic pastures. We compared the efficacy of four distinct kinds of bird detection (auditory, visual, auditory followed by visual, and visual followed by auditory) in relation to the habitats. Visual and auditory detection were proportional and compensatory considering all habitats surveyed. The results suggest that point counts can be efficiently used for bird surveys in local agroecosystems, where the habitats variability allows the balance between visual and auditory detections.

M. Penteado (✉)
Instituto Chico Mendes de Conservação da Biodiversidade (ICMBio),
Av. Manoel Hipólito do Rêgo, 1907, São Sebastião, SP 11600-00, Brazil
e-mail: marli.penteado@icmbio.gov.br

W. R. Silva
Department of Zoology, Biology Institute, Universidade de Campinas,
Caixa Postal 6109, Campinas, SP 13083-970, Brazil
e-mail: wesley@unicamp.br

L. M. Verdade
Centro de Energia Nuclear na Agricultura, Universidade de São Paulo,
Caixa Postal 96, Piracicaba, SP 13416-000, Brazil
e-mail: lmverdade@usp.br

L. M. Verdade et al. (eds.), *Applied Ecology and Human Dimensions in Biological Conservation*, DOI: 10.1007/978-3-642-54751-5_10,

10.1 Introduction

The precision and accuracy of the available methods of bird survey can be considerably affected by the species in question, its habitat, and the observer's skills (Lack 1937; Ralph and Scott 1981; Verner 1985; Bibby et al. 1992; Casagrande and Beissinger 1997; Jones et al. 2000; Simons et al. 2007; Stanislav et al. 2010). Therefore, such methods should be evaluated before used on a combination of species/habitat they have not been developed for (Karr 1981; Scott and Ralph 1981; Granholm 1983; Verner 1985; Verner and Ritter 1986).

It can be particularly difficult to survey birds in agroecosystems because agricultural landscapes can be formed by a heterogeneous mosaic with different kinds of land use. In such circumstance, each patch of the mosaic can present distinct patterns of vegetation structure (both vertical and horizontal), as well as presence of humans and livestock, pesticides, and different levels of edge effects. These local and regional landscape characteristics can affect birds' detectability in a species-specific way (Oelke 1981) what can influence, by its turn, bird survey methods in distinct ways (Bibby et al. 1992; Ralph et al. 1995).

Point counts is currently the most used method of bird survey in ecological studies possibly because it can easily fit in experimental design, generating independent sampling unities; and, it usually detects more bird species than other methods (Blondel et al. 1981; Reynolds et al. 1980; Edwards et al. 1981; Ralph 1985; Szaro and Jakle 1985; Bibby et al. 1992). Point counts can be considered as a transect line with null length and speed (Bibby et al. 1992), and this is possibly the reason for the larger number of total and rare species detected by point counts in relation to transects (Edwards et al. 1981; van Ripper III 1981). Point counts are particularly convenient for forested habitats where identifying species while walking through dense vegetation can be rather difficult (van Ripper III 1981; Verner 1985; Ralph et al. 1995). However, point counts can be particularly adequate for bird community studies in mosaics where habitat characteristics can be related to the occurrence of individual species (Oelke 1981; Bibby et al. 1992; Hvenegaar 2011). In such circumstance, the occurrence of replicates on the landscape mosaic can allow the use of statistical tests for the hypothetical habitat-species relationships (Verhulst et al. 2004; Woodhouse et al. 2005). For this reason, point counts have been recently used in comparative studies between open and forested habitats in agricultural landscapes (Cárdenas et al. 2003; Verhulst et al. 2004; Harvey et al. 2005; Moreira et al. 2005; Woodhouse et al. 2005). However, the efficiency of the method in such circumstances has not yet been evaluated. This is the main goal of the present study. In order to do so, we compared open and forested habitats of an agricultural landscape of the state of São Paulo, Southeastern Brazil, in terms of the frequency of bird detections by visual and auditory contacts using point counts.

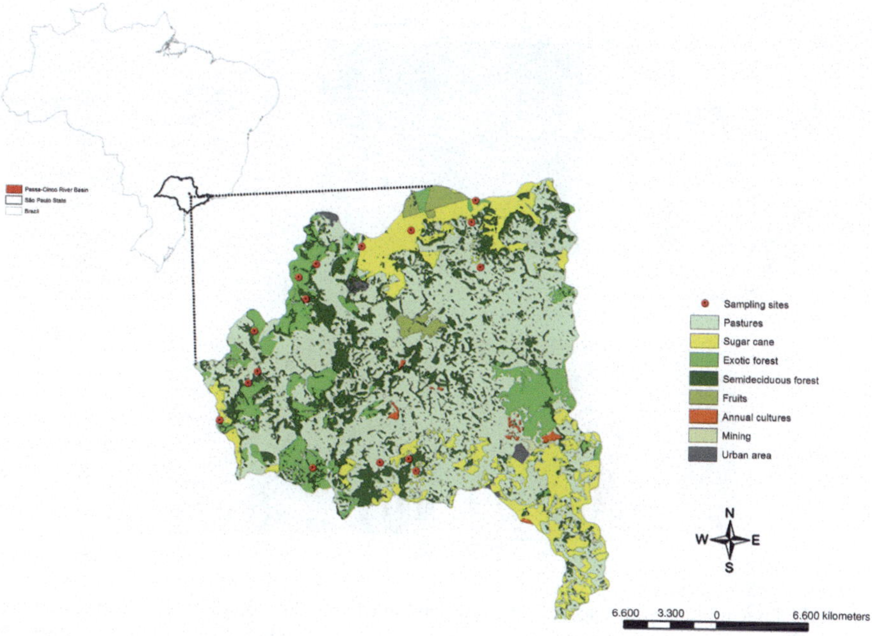

Fig. 10.1 Location of this study sites in Passa-Cinco river basin, Central-Eastern region of the state of São Paulo, Southeastern Brazil

10.2 Study Area, Sampling Design, and Statistical Analyses

This study was carried out at the Passa-Cinco river basin (22°05′–22°30′S, 47°30′–47°30′W) in the Central-eastern region of the state of São Paulo, Southeastern Brazil (Fig. 10.1). This river basin spreads over an area of 280 km^2 and is covered by exotic pastures (51.7 %), sugarcane plantations (14.1 %), *Eucalyptus* plantations (10.8 %), and remnant fragments of semi-deciduous Atlantic forest (15.6 %) (Valente 2001). Deforestation began to take place in this region on the eighteenth century for agricultural purposes (Dean 1977). Nowadays, it comprises an agricultural landscape with more or less isolated native forest fragments and urban developments that well represent the countryside of the state of São Paulo in Southeastern Brazil (Rodrigues 1999).

In this study, we sampled bird species by point counts in four kinds of landscape "attributes" (sensu Forman 1995) that comprise more than 90 % of the total area of the Passa-Cinco river basin, as mentioned above: fragments of semi-deciduous Atlantic forest, *Eucalyptus* plantations, sugarcane plantation, and exotic pastures (mostly *Brachiaria* spp). Four replicates of each of these landscape attributes were used totalizing 16 sites spaciously distributed on a nested way (sensu Zar 1999:303) (Fig. 10.2).

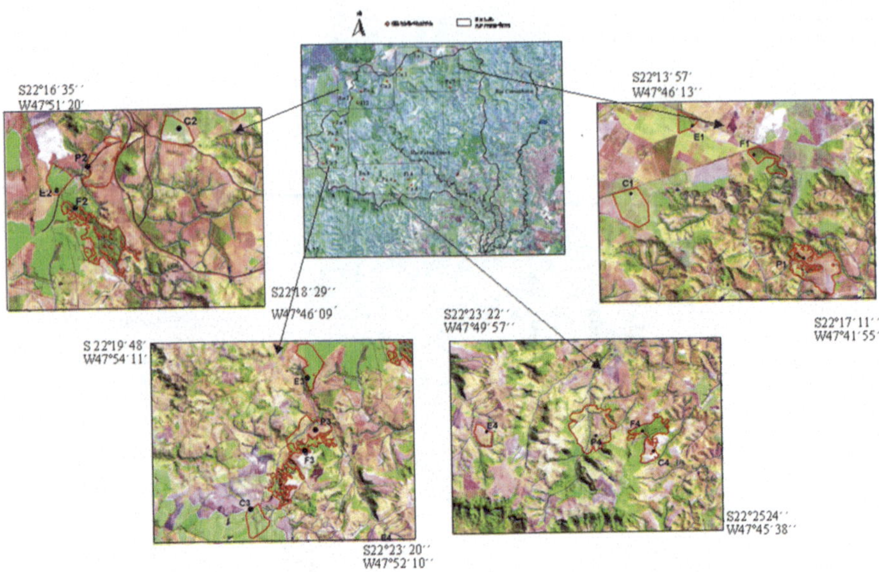

Fig. 10.2 Four replicates of each of the landscape attributes (i.e., fragments of semi-deciduous Atlantic florest, *Eucalyptus plantations*, sugarcane plantations and exotic pastures), totalizing 16 sites spaciously distributed on a nested way in Passa-Cinco River basin, Central-eastern region of the state of São Paulo Southeastern Brasil

We carried out 11 surveys per study site, six at the rainy season (October–March) and five at the dry season (April–September), from September 2003 to January 2005, totalizing 176 surveys. We used five points per site with a distance of 200 m from each other at the core area of the patch (200 m of minimum distance from the edge). Surveys started 30 min after sunrise and were carried out by a single observer (MP) during the whole study. The visiting time per point was 10 min. Four kinds of detection were considered, as follows: vocalization (voc), visualization (vis), vocalization followed by visualization (voc/vis), and visualization followed by vocalization (vis/voc). All birds detected were considered regardless the distance from the observer unless they were out of the patch or flying over it (Blondel et al. 1981). A directional microphone Sennheiser® System K6-ME 67 and a tape recorder Sony® DAT TCD-D100 were used for vocalization recording. A binocular Zeiss® Deltrinten 8 × 30 was used to help visualizations.

The species incidence curve for the whole study area has been fit to an asymptotic model by nonparametric Bootstrap procedure in EstimateS Win 7.0 (Colwell 2004). We used Kruskal-Wallis test to compare landscape attributes (considering 16 study sites) in terms of the frequency of occurrence of the distinct detection types, as Levene's Test rejected homoscedasticity for both visualization and vocalization datasets (LS = 16.11, df = 172, $p < 0.001$, LS = 13.99, df = 172, $p < 0.001$, respectively). We then compared means by post hoc Tukey HSD and Duncan tests (Zar 1999).

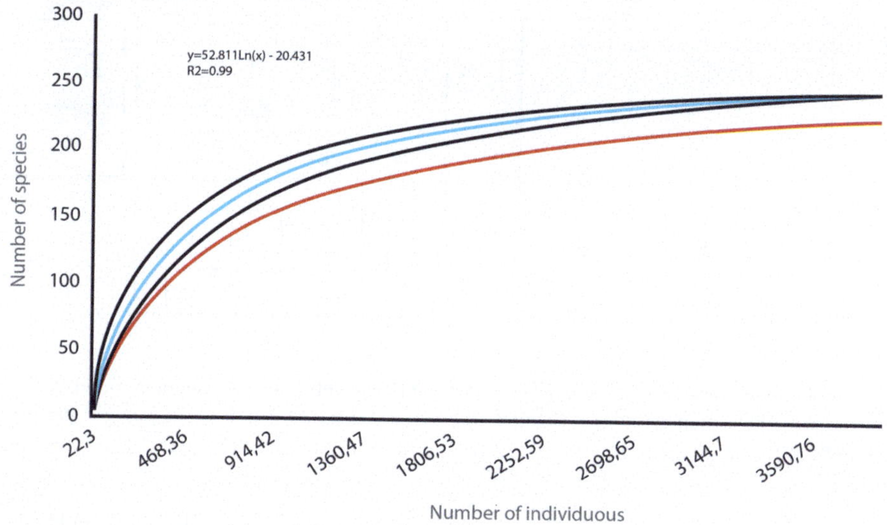

Fig. 10.3 Species incidence curve for the whole study area estimated by nonparametric bootstrap procedure (EstimateS win 700)

Table 10.1 Number of bird detections per landscape attribute

Landscape attribute detection type	Native forest	Eucalyptus	Pasture	Sugarcane	Total
Visualization	229	186	605	241	1,261
Vocalization	653	237	280	157	1327
Visualization/vocalization	108	56	42	49	255
Vocalization/visualization	141	63	244	38	486
Total	1,131	542	1,171	485	3,329

In order to test for possible variation in different landscape attributes, we took two different approaches: (a) we considered each survey per site as our sampling unit and compared landscape attribute by Kruskal-Wallis and post hoc Tukey and Duncan tests; and (b) for the sugarcane plantations (where vegetation height varies from 0 to 2.5 m), we used t-test to compare initial growth and pre-harvest stage for species richness and total abundance.

10.3 Results and Discussion

A total number of 224 species have been detected for the whole study area, which is compatible with the estimated asymptotic model of the species incidence curve (approximately 250 species) (Fig. 10.3). This suggests that the sampling effort was enough to detect all species present in the area.

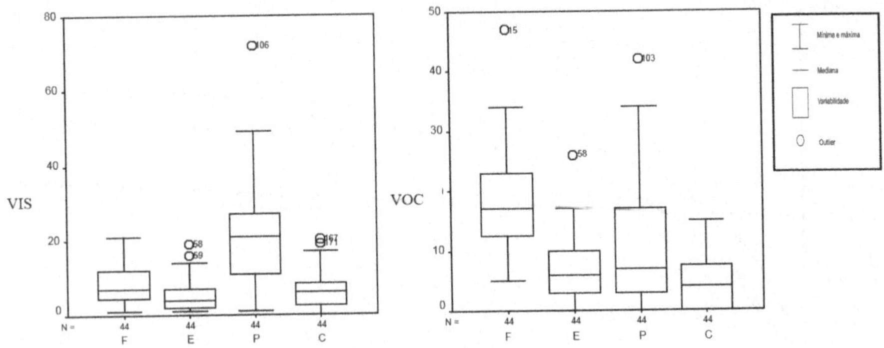

Fig. 10.4 Total number of bird detection per landscape attribute (*VIS* number of visualization reports, *VOC* number of vocalization reports, *F* native forest, *E eucalyptus*, *P* pasture, *C* sugarcane)

A total number of 3,329 records of birds have been done, being 1,327 voc, 1,261 vis, 486 voc/vis, and 255 vis/voc (Table 10.1). Most of the records were done in exotic pastures (35.2 %, $N = 1{,}171$), followed by native forest fragments (34.0 %, $N = 1{,}131$), *Eucalyptus* plantations (16.3 %, $N = 542$), and sugarcane plantations (14.5 %, $N = 485$) (Table 10.1). If congregated, primarily visual (i.e., vis + vis/voc) represented 45.5 % ($N = 1516$) and primarily auditory (i.e., voc + voc/vis) represented 54.5 % ($N = 1813$) of the total records. There were only five outliers for visualizations and three for vocalizations (Fig. 10.4).

There was a significant variation among landscape attributes in terms of the frequency of occurrence of visual and auditory birds' detection (Kruskal-Wallis: $H = 61.32$, $df = 15$, $p < 0.001$, $H = 66.48$, $df = 15$, $p < 0.001$, respectively, for visualizations and vocalizations). Considering the spatial variation of the present study (i.e., taking each site as sampling unity, $n = 16$), visualizations were more frequent in pastures, whereas vocalizations were more frequent both in pastures and forest habitat (native forest for Tukey and *Eucalyptus* for Duncan tests) (Table 10.2). Considering each sampling per site as sampling unity ($n = 44$ in each habitat), visualizations were also more frequent in pastures, vocalizations were more frequent in native forest, and mixed detection types (i.e., vis/voc and voc/vis) were more frequent in pastures and native forest fragments (with the exception of vis/voc for Duncan test, in which only pastures were kept apart). On the other hand, *Eucalyptus* and sugarcane plantation presented similar patterns of birds' detection along the year, whereas native forest and pastures present different patterns for vocalizations and visualizations (Table 10.2).

Sugarcane plantations did not present differences both in terms of species richness and abundance comparing early and late agricultural stages ($t = 0.305$, $df = 228$, $p = 0.760$ for abundance). However, as mentioned above, both plant biomass and height dramatically vary along the year from plantation to harvest.

Considering the whole study area, there was no difference in terms of the frequency of occurrence between visual and auditory bird detection. On the

Table 10.2 Post hoc Tukey and Duncan tests for bird detection by landscape attribute

Detection type	Test	N	a	b	c
VIS	Tukey	16	Fl, E, C	P	
	Duncan	16	Fl, E, C	P	
VOC	Tukey	16	E, P, C	Fl, P	
	Duncan	16	E, C	E, P	Fl
VIS	Tukey	44	Fl, E, C	P	
	Duncan	44	Fl, E, C	P	
VOC	Tukey	44	E, C	P	Fl
	Duncan	44	E, C	P	Fl
VIS	Tukey	44	Fl, E, C	P	
	Duncan	44	Fl, E, C	P	
VOC	Tukey	44	E, P, C	Fl	
	Duncan	44	E, P, C	Fl	
VIS/VOC	Tukey	44	Fl, E, C	Fl, P	
	Duncan	44	Fl, E, C	P	
VOC/VIS	Tukey	44	Fl, E, C	Fl, P	
	Duncan	44	Fl, E, C	Fl, P	

For spatial variation, each site ($N = 16$) was considered as sampling unity. For temporal variation, each sampling per site ($N = 44$) was considered as sampling unity (visualization *VIS*, vocalization *VOC*; visualization followed by vocalization *VIS/VOC*, vocalization followed by visualization *VOC/VIS*; native forest *Fl*, eucalyptus *E*, pasture *P*, sugarcane *C*; a < b < c)

other hand, considering different landscape attributes, there seems to be some compensation between visual and auditory detection in such a way that different environments can be equally surveyed by point counts method. In open habitats such as pastures and the early stages of sugarcane plantations, visualizations are predominant. On the contrary, in densely vegetated forest habitats (either native or exotic), vocalization records occur more often. This means that point counts can be an efficient method of bird survey independently of the vegetation structure, although it has been created originally for forest habitats.

The variability found in this study can be at least partially explained by two factors: temporal (most seasonal) variation in vegetation biomass (more dramatic in sugarcane plantations, but also relevant for the others), and spatial (mostly microgeographic) variation in patch structure, history, and agricultural management (more dramatic in pastures where different areas can have a distinct number of remaining trees, but also relevant for the others). There is an obvious temporal effect at the later, but in a larger scale in relation to the former, what can be assumed as just spatial variation in a short-term study such as this, for the sake of simplicity.

The detectability of birds by both visual and auditory contacts can be affected by vegetation, topography, and the distance from the observer (Oelke 1981). In forest habitats, these factors can be even more relevant than the period of the day during which the surveys are carried out (Richards 1981). This pattern can be noted when surveys are carried out in standardized periods (e.g., always early in the morning). The vegetation structure in agroecosystems is different from pristine forested areas. This results in a certain similarity between forest habitats (i.e., native and *Eucalyptus* plantations) and open habitats (i.e., pastures and sugarcane plantations) in terms of bird detection.

Agricultural landscapes are currently widespread. Conservation of bird species can be possibly no longer assured only by conservation areas such as national parks and biological reserves (Green et al. 2005; Mulwa et al. 2012). In such context, bird surveys in agricultural landscapes are urged in two ways. By the species detected on surveys, we might infer about the actual relevance of agricultural landscapes for bird conservation. By the species nondetected on surveys, we might establish and experimentally test hypotheses related to possible causes of local extinctions and population declines, most of them possibly related to agricultural practices. To be effective in doing so, we should choose adequate survey methods in which "omission errors" (Fielding and Bell 1997) are least, i.e., nondetected species are as close as possible to the actually absent ones. Point counts seem to attend this assumption for agricultural landscapes of the state of São Paulo in southeastern Brazil.

Acknowledgments We would like to thank Carlos Yamashita for suggestions and precious comments along this study. We also would like to thank Edson Davanzo, Henrique Rocha, and Roberto Nogueira for the invaluable help in the field. The present study is part of a multi-taxa survey project supported by IBAMA and the Biota Program /FAPESP (Proc. No. 01/13251-4).

References

Bibby CJ, Burgess ND, Hill DA (1992) Bird census techniques. London Academic Press, London

Blondel J, Ferry C, Fronchot B (1981) Point counts with unlimited distance. In: Ralph CJ, Scott JM (eds) Estimating numbers of terrestrial birds: studies in avian biology. Allen Press, Kansas, pp 414–420

Cárdenas G, Harvey CA, Ibrahim M, Finergan B (2003) Diversidad y riqueza de aves en diferentes hábitats en paisaje fragmentado en Cañas, Costa Rica. Agroforesteria en las Am 10:78–85

Casagrande DG, Beissinger SR (1997) Evaluation of four methods for estimating parrot populations size. Condor 99:445–457

Colwell RK (2004) EstimateS version 7.0: statistical estimation of species richness and shared species from samples New York

Dean W (1977) Rio Claro: Um sistema brasileiro e grande lavoura—1820–1920. Paz e Terra Rio de Janeiro

Edwards DK, Dorsey GL, Crawford JA (1981) A comparison of three avian census methods. In: Ralph CJ, Scott JM (eds) Estimating numbers of terrestrial birds: studies in avian biology. Allen Press, Kansas, pp 170–176

Fielding AH, Bell JF (1997) A review of methods for the assessment of prediction errors in conservation presence/absence models. Environ Conserv 24:38–49

Forman RTT (1995) Land mosaics: the ecology of landscapes and regions. Cambridge University Press, Cambridge, UK

Granholm SL (1983) Bias in density estimates due to movement of bird. Condor 85:243–248

Green RE, Cornell SJ, Scharlemann JPW, Balmford A (2005) Farming and the fate of wild nature. Science 307:550–555

Harvey CA, Villanueva C, Villacis J, Chacon M, Muñoz D, López M, Ibrahim M, Gómez R, Taylor R, Martinez J, Navas A, Saenz J, Sánchez D, Medina A, Vilchez S, Hernández B, Perez A, Ruiz F, López F, Lang I, Sinclair FL (2005) Contribution of live fences to ecological integrity of agricultural landscapes. Agricult Ecosyst Environ 3:200–230

Hvenegaar GT (2011) Validating bird diversity indicators on farmland in east-central Alberta, Canada. Ecolog Indic 11:741–744

Jones J, Mcleish WJ, Robertson RJ (2000) Density influences census technique accuracy for Cerulean Warblers in Eastern Ontario. J Field Ornithol 71:46–56

Karr JR (1981) Surveying birds in the tropics. In: Ralph CJ, Scott JM (eds) Estimating numbers of terrestrial birds: studies in avian biology. Allen Press, Kansas

Lack D (1937) A review of bird census work and bird populations problems. Ibis 14:369–395

Moreira F, Beja P, Morgado R, Reino L, Gordinho L, Delgado A, Borralho R (2005) Effects of field management and landscape context on grassland wintering birds in Southern Portugal. Agricult Ecosyst Environ 109:59–74

Mulwa RK, Böhning-Gaese K, Schleuning M (2012) High bird species diversity in structurally heterogeneous farmland in Western Kenya. Biotropica 44:801–809. doi:10.1111/j.1744-7429

Oelke H (1981) Limitations of estimating bird populations because of vegetation structure and composition. In: Ralph CJ, Scott JM (eds) Estimating numbers of terrestrial birds: studies in avian biology. Allen Press, Kansas, pp 316–321

Ralph CJ (1985) Habitat association patterns of forest and steppe birds of Northern Patagonia, Argentina. Condor 87:471–482

Ralph CJ, Scott JM (eds) (1981) Estimating numbers of terrestrial birds: studies in avian biology. Allen Press, Kansas, USA

Ralph CJ, Droege S, Sauer J (1995) Managing and monitoring birds using point counts: standards and applications. In: Ralph CJ, Sauer JR, Droege S (eds) Monitoring bird populations by point count. Forest Service General Technical Report, USDA, pp 25–34

Reynolds RT, Scott JM, Nussbaun RA (1980) A variable circular plot method for estimating bird numbers. Condor 82:309–313

Richards DG (1981) Environmental acoustic and censuses of singing birds. In: Ralph CJ, Scott JM (eds) Estimating numbers of terrestrial birds: studies in avian biology. Allen Press, Kansas, pp 297–300

Rodrigues RR (1999) A vegetação de piracicaba e municípios de entorno. Circular técnica IPEF 189:1–18

Scott JM, Ralph CJ (1981) Estimating birds: introduction. In: Ralph CJ, Scott JM (eds) Estimating numbers of terrestrial birds: studies in avian biology. Allen Press, Kansas, pp 1–2

Simons TR, Alldredge MW, Pollock KH, Wettroth JM (2007) Experimental analysis of the auditory detection process on avian point counts. Auk 124:986–999

Stanislav SJ, Pollock KH, Simons TR, Alldredge MW (2010) Separation of availability and perception processes for aural detection in avian point counts: a combined multiple-observer and time-of-detection approach. Avian Conserv Ecol 5:3–13 (online) URL: http://www.ace-eco.org/vol5/iss1/art3/

Szaro RC, Jakle MD (1985) Avian use of a desert riparian island and its adjacent scrub habitat. Condor 87:511–519

Valente ROA (2001) Análise da estrutura da paisagem na bacia do rio Corumbataí, SP. Piracicaba. 2001. 144 p. Dissertação (Mestrado em Recursos Florestais)—Escola Superior de Agricultura "Luiz de Queiroz", Universidade de São Paulo, Piracicaba

Van Riper C III (1981) Comparison of methods for estimating numbers of terrestrial birds. In: Ralph CJ, Scott JM (eds) Estimating numbers of terrestrial birds: studies in avian biology. Allen Press, Kansas, pp 217–218

Verhulst J, Báldi A, Kleijn D (2004) Relationship between land-use intensity and species richness and abundance of birds in Hungary. Agricult Ecosys Environ 104:465–473

Verner J (1985) Assessement of counting techniques. Cur Ornithol 2:247–302

Verner J, Ritter LV (1986) Hourly variation in morning point counts of birds. Auk 103:117–124
Woodhouse SP, Good JEG, Lovett AA, Fuller RJ, Dolman PM (2005) Effects of land-use and agricultural management on birds of marginal farmland: a case study in the Llyn peninsula, Wales. Agricult Ecosys Environ 107:331–340
Zar ZH (1999) Biostatistical analysis. Prentice Hall, Englewood Cliffs, p 718

Chapter 11
The Use of Stable Isotopes Analysis in Wildlife Studies

Thiago S. Marques, Neliton R. F. Lara, Plínio B. Camargo, Luciano M. Verdade and Luiz A. Martinelli

Abstract The application of stable isotopes analysis in wildlife studies has increased in recent decades due to the wide range of information that can be obtained with this methodology. This chapter aims to present the basic principles of the stable isotopes analysis and their potential applications in wildlife studies. The main topics presented are diet reconstruction, trophic level, animal movements, tissue turnover rates, and ecotoxicology.

11.1 Introduction

Stable isotopes analysis show a wide range of applications in biological, earth, and environmental sciences. However, in recent decades wildlife studies using this methodology increased exponentially promoting a major development in this area of knowledge (Gannes et al. 1997, 1998; Crawford et al. 2008; Layman et al. 2012). This is probably due to the large amount of information that can be obtained through this methodology to answer different types of questions related to wild animals.

T. S. Marques (✉) · N. R. F. Lara · P. B. Camargo · L. A. Martinelli
Laboratório de Ecologia Isotópica—Centro de Energia Nuclear na Agricultura, Universidade de São Paulo, PO Box 96, Piracicaba, SP 13400-970, Brazil
e-mail: thiagomq@yahoo.com.br

N. R. F. Lara
e-mail: neliton.lara@gmail.com

P. B. Camargo
e-mail: pcamargo@cena.usp.br

L. A. Martinelli
e-mail: zebu@cena.usp.br

L. M. Verdade
Centro de Energia Nuclear na Agricultura, Universidade de São Paulo, Caixa Postal 96, Piracicaba, SP 13416-000, Brazil
e-mail: lmverdade@usp.br

L. M. Verdade et al. (eds.), *Applied Ecology and Human Dimensions in Biological Conservation*, DOI: 10.1007/978-3-642-54751-5_11,

The use of stable isotopes is based on the fact that the isotopic compositions vary in a predictable way, as the element moves through the various ecosystem compartments (Martinelli et al. 2009). Many chemical processes result in isotopic fractionation, because of the mass difference between the light and heavy isotopes. Due to these characteristics, stable isotopes can be used as a biological tracer in ecological studies.

The analysis of stable isotopes provides a clear advantage in identifying differences in resources use at different scales (Dalerum and Angerbjörn 2005) allowing the assessment of long-term ecological trends, needed to management and conservation plans for wild species. Furthermore, new quantitative analytical approaches have emerged to elucidate various aspects of the biology of wild species based on the stable isotopes composition increasing its potential applicability (Layman et al. 2012).

The conservation biology deals with the causes and consequences of biodiversity loss. In this context, the development of both technological tools and conceptual basis is necessary to perceive, identify, and solve problems. This chapter has the purpose of showing how the stable isotopes tool can be used to answer ecological questions.

This chapter provides the presentation of main stable isotope analysis applications in wildlife studies comprising diet reconstruction, trophic level, animal movements, tissue turnover rates, and ecotoxicology. The themes breadth and the rapid growth of studies using this methodology make unfeasible a great depth and presentation of all published studies for each application. Therefore, our goal is to present the topics in a simple and concise form, wherever possible, providing examples of studies conducted in tropical environments. Our intention is that this work will serve as a guide for researchers wishing to get to know the applications of isotope methodology in wildlife studies.

Box 1: Stable Isotope Methodology

The study of applications of stable isotopes analysis in wildlife studies requires a brief review of some basic concepts discussed below. Isotopes are species of the same chemical element that have different atomic masses. This is due to variations in the neutrons number in the nucleus, for example, nitrogen isotope ^{14}N presents atomic mass equal to 14 (7 protons + 7 neutrons), and ^{15}N presents atomic mass equal to 15 (7 protons + 8 neutrons). In addition, the isotopes are considered stable when they do not undergo radioactive decay, thus maintaining the same mass over time.

Stable isotopes have different natural abundances being lighter isotopes (lower atomic mass) more abundant, while heavier isotopes (higher atomic mass) are less abundant. These differences in the isotopes concentrations can be measured using a mass spectrometer.

The stable isotope ratios (heavy/light, e.g., $^{13}C/^{12}C$, $^{15}N/^{14}N$ $^{2}H/^{1}H$) are usually expressed by the notation (δ) and are related to international standards (Fry 2006). The δ values are numerically small (of the order of 10^{-2}), so the results of these expressions are usually multiplied by 1,000, being referred to as parts per mil (‰). These values can be either positive or negative, depending on the isotopes ratios. International standards have been defined for each of the elements. For example, the carbon standard is Peedee Belamite (PDB), a Belemnnitella fossil of the Peedee formation in the South Carolina (USA), nitrogen standard is air (N_2) and hydrogen standard is Vienna Standard Mean Ocean Water (VSMOW).

11.2 Uses in Wildlife Studies

11.2.1 Diet Reconstruction

Conventional methods of direct observation, stomach contents, and feces analysis have been traditionally used in ecological studies in order to understand individuals' diet (Litvaitis 2000). These methodologies provide important information about the recent diet; however, biases associated with different levels of items digestibility and accidental ingestion are common (Martinelli et al. 2009). In this context, the application of stable isotopes analysis to reconstruct the diet has gained more prominence in recent years (Boecklen et al. 2011).

The use of this methodology is based on the fact that the isotopic composition of animal tissues reflect the isotopic composition of their diet discounted the isotopic fractionation between diet–animals (DeNiro and Epstein 1978), so there is the possibility to track the diet assimilated by the animal (Ramos and González-Solís 2012). The stable isotopes analysis provide valuable ecological information in situations that conventional methods are disabled or as complementary to classical studies of stomach content analysis, for example, the investigation of the resources partition from individual to community level (Inguer and Bearhop 2008).

Some important aspects about isotopic analyzes should be highlighted before utilization. The isotopic composition of food sources must be distinct (e.g., C_3 or C_4 plants) and the animal tissue analyzed must be chosen appropriately taking into account the study objectives, the turnover rate and the isotopic fractionation (Gannes et al. 1998). When these aspects are attended, the isotopic compositions of consumers and their potential prey can be used for a qualitative or quantitative analysis of diet reconstruction (Layman et al. 2012).

Currently, there are many mixing models that yield relative contributions of several diet sources providing benefits in comparison with traditional methods for diet analysis (Phillips and Gregg 2003; Parnell et al. 2010, 2013; Boecklen et al. 2011; Erhardt and Bedrick 2013). The stable isotope analysis in R (SIAR), (Parnell et al. 2010) has been one of the most common models presently used by researchers in wildlife studies. This model utilizes linear equations and Bayesian statistical techniques to report ranges of proportional source pool contributions to consumers. Its advantage over previous models is incorporate uncertainty and variation in input parameters. Several examples of the isotope methodology application for diet studies can be found in the literature.

Oliveira (2006) studied seasonality of energy sources of tambaqui (*Colossoma macropomum*), fish of great economic importance in the floodplains of the Amazon region and that exhibit diverse feeding habitats, which can only be accessed by the combination of classical analysis of stomach contents and stable isotope techniques. The analysis of stomach contents showed that the relative importance of the food items varied with water level (rising, high, falling, and low). Fruits and seeds become available during periods of high water level when tambaqui have access to the flooded forest. However, lakes are disconnected from the rivers during the low water period, making the availability of food resources different. The $\delta^{13}C$ and $\delta^{15}N$ values of fish tissues also varied during the hydrological cycle. C_3 plant material (including fruits and seeds) was the main contributor to the tambaqui biomass with values between 55 and 95 % depending on the water level. C_4 plants have little importance in the tambaqui diet (maximum contribution = 26 %) probably because of its low nutritional value. Zooplankton played a role in supplying nitrogen to tambaqui.

Lara et al. (2012) studied the trophic relationship and primary carbon sources of diets between two sympatric freshwater turtles widely distributed in the Amazon basin, *Podocnemis unifilis* and *Podocnemis expansa*, using carbon and nitrogen stable isotopes analysis. No differences were found between the two species in relation to $\delta^{13}C$ (means *P. unifilis*: −26.2 ‰; *P. expansa*: −26.1 ‰), but *P. unifilis* had higher values of $\delta^{15}N$ than *P. expansa* (means *P. unifilis*: 7.6 ‰; *P. expansa*: 5.1 ‰), indicating a possible trophic change due to exploitation of different food resources. In addition, the values of $\delta^{13}C$ show the dependence of these species on C_3 plants, which represent their main source of basal energy. These two species of freshwater turtles have a diet based on aquatic plants, algae, seeds, leaves, fruits, flowers, roots, stems, and occasionally small animals (Pritchard and Trebbau 1984). However, migrations to small lakes made by *P. unifilis* during the flood season enables the exploitation of a broader range of food resources that *P. expansa*.

Marques et al. (2013) studied the intraspecific isotopic niche variation in Broad-snouted caiman (*Caiman latirostris*) in a silvicultural landscape in Brazil through the use of carbon and nitrogen stable isotopes. Discrete ontogenetic variations in the isotopic niche and sexual difference only for juveniles were identified analyzing claw samples collected from juveniles, adults, and hatchlings of *C. latirostris*. There is a progressive increase in stable isotope compositions values ($\delta^{15}N$ and $\delta^{13}C$) in relationships to animals' snout–vent length. These results may indicate differences in the exploitation of diet resources to decrease intraspecific competition. Crocodilians show a dramatic increase in body mass during ontogenetic development, which can result in diet shift from invertebrate to vertebrate items. Dietary studies using stomach contents suggest that the species can exploit a wide variety of prey such as insects, arachnids, crustaceans, snails, fish, amphibians, reptiles, birds, and mammals (Melo 2002; Borteiro et al. 2009).

Furthermore, the stable isotopes methodology has also been applied to the paleodiet reconstruction using fossil samples (Parkington 1991; Pate 1997; MacFadden 2000; Koch 2007; Clementz 2012). This methodology has opened a new perspective to become an important tool for paleontologists infer the diet of extinct animals because direct observation is not possible. The collagen (protein of bone tissue) removed from bones and teeth are the material commonly used for these studies because remain preserved even with the passage of time (DeNiro 1987). The preservation quality of original isotopic information in this material can be evaluated by ratio of carbon to nitrogen (C:N) in samples (Ambrose 1990; van Klinken 1999). The C/N values should be between 2 and 3, so it can be sure that there was no contamination from exogenous sources (DeNiro 1985; Martinelli et al. 2009).

The power of isotopic tool in paleodiet reconstruction can be exemplified through the study of MacFadden et al. (1999) with six sympatric horses of 5 million years old (late Hemphillian) from fossil deposits of Florida. Traditional morphological studies of tooth crown height indicate that these animals with high-crowned teeth have fed on abrasive grasses, but enamel $\delta^{13}C$ values in combination with tooth microwear data indicate that these horses in this study were not exclusive C_4 grazers but also included mixed feeders and C_3 browsers. C_4 plants in this context include most grasses, while C_3 plants include most leafy, woody, and other soft plants (browse). Therefore, this study demonstrated that horses can partition their food resources from almost pure C_4 grazers to principally C_3 browsers, contrary to previous studies with others approaches.

11.2.2 Trophic Level

The stable isotopes analysis are extremely useful for studies about nutrients and energy transfer in food webs. The nitrogen stable isotopes are often used in trophic web studies due to the expected increase in ^{15}N over successive levels, according consumers tissues are enriched relative to its diet (Kelly 2000; Fry 2006). The consumer tissues have differents $\delta^{15}N$ values due to assimilation and excretion of nitrogen (Macko et al. 1986; Olive et al. 2003), with the excretion of lighter nitrogen (^{14}N) in the urine. This preferential removal of ^{14}N amine groups occurs by the enzymes responsible by desamination and transamination of aminoacids (Macko et al. 1986, 1987).

In this context, the trophic position of an animal can be estimated based on the $\delta^{15}N$ values of the food chain and on the ^{15}N enrichment values in each trophic level (Post 2002). The transfer of trophic level varies on average 2.5 ‰ (Fry 1991) to 3.4 ‰ for $\delta^{15}N$ (DeNiro and Epstein 1981; Minagawa and Wada 1984). However, these values can vary according to the number of trophic transfers. In general, 3.4 ‰ refers to calculations of trophic multiple paths (Post 2002), whereas values for a single transfer trophic may vary between 2 and 5 ‰ (Adams and Sterner 2000; McCutchan et al. 2003). Furthermore, trophic level of consumer can be estimated applying the formula adapted from Vander Zanden et al. (1997):

$$\mathrm{TP} = \left(\frac{\delta^{15}\mathrm{N}_{\mathrm{consumer}} - \delta^{15}\mathrm{N}_{\mathrm{baseline}}}{\Delta\delta^{15}\mathrm{N}} \right) + \lambda$$

where *TP* is the trophic position of the consumer, $\delta^{15}N_{consumer}$ is the nitrogen isotopic value of the consumer, $\delta^{15}N_{baseline}$ is the mean nitrogen isotopic value of the base of the food chain assumed (i.e., primary producers), $\Delta\delta^{15}N$ is the "enrichment factor," and $\lambda =$ is the trophic position of the organism used to estimated $\delta^{15}N$ baseline.

This method is highly dependent on the generation of the suitable base isotopic representing the spatial and temporal variation of $\delta^{15}N$ within and between systems of interest (Post 2002). Therefore, it depends on a good estimate of the isotopic values on the lower trophic level of the system and the resources used by consumers. In addition, the use of this methodology depends on the estimation of discrimination factors ($\Delta^{15}N$) between tissues and diet (Caut et al. 2009). Discrimination factors show several sources of variation, like food type, physiological stress, lipid extraction, diet quality, taxa, and tissues (Hobson et al. 1993; McCutchan et al. 2003; Roth and Hobson 2000; Caut et al. 2009).

The body condition and consequent metabolic state also may affect the fractionation in the organisms. Animals in a starvation state show a progressive enrichment in $^{15}N/^{14}N$ rate, in a similar process to what happens along the trophic chain (Hobson et al. 1993). In this case, ^{14}N excreted is not replaced by the protein diet, so the animal becomes progressively enriched in ^{15}N as its hunger state increases. Therefore, the $\delta^{15}N$ can also be used as an indicator of changes in body condition (Hobson et al. 1993).

Manetta et al. (2003) used stomach contents and stable isotopes composition of nitrogen ($\delta^{15}N$) to verify the trophic position (TP) of the main species of fishes, of the Paraná River floodplain, Brazil. There was no difference between both methods and indicate that *Loricariichthys platymetopon* (TP by stomach contents: 2.0; TP by stable isotope: 2.1), *Schizodon borellii* (TP by stomach contents: 2.0; TP by stable isotope: 2.4), *Leporinus lacustris* (TP by stomach contents: 2.1; TP by stable isotope: 2.7), and *L. friderici* (TP by stomach contents: 2.0; TP by stable isotope: 2.3) are primary consumers and *Auchenipterus osteomystax* (TP by stomach contents: 3.5; TP by stable isotope: 3.8), *Iheringichthys labrosus* (TP by stomach contents: 3.0; TP by stable isotope: 3.6), and *Serrasalmus marginatus* (TP by stomach contents: 3.9; TP by stable isotope: 3.5) are secondary consumers. A great intraespecific variability of $\delta^{15}N$ was found in several fish species, for example, *I. labrosus* (omnivorous) possibly as a result of great diversity of food items in its diet, including higher plants, detritus, besides prey from different trophic levels. The high plasticity of food itens in fish species may mean that changes in the trophic hierarchy can occur depending on environmental conditions.

Estrada et al. (2003) estimated the trophic positions of the blue shark (*Prionace glauca*), shortfin mako (*Isurus oxyrinchus*), thresher shark (*Alopias vulpinus*), and basking shark (*Cetorhinus maximus*) from Atlantic Ocean near to Martha's Vineyard island, USA using stable isotope ratios of nitrogen ($\delta^{15}N$). Sharks are apex predators in the marine environment and their feeding ecology can affect the community structure. The basking shark had the lowest trophic positions (3.1) followed in crescent order by blue shark (3.8), shortfin mako (4.0), and thresher shark (4.5). Trophic position of sharks is closely related to the exploitation of food resources, for example, basking shark known to feed solely on zooplankton, comparisons with isotopic values of prey species suggest that blue shark and shortfin mako forage primarily on fish prey and thresher shark feed mainly on cephalopods.

11.2.3 Animal Movements

Traditional radiotelemetry techniques have been used to detect movement patterns in wild animals (Millspaugh and Marzluff 2001; Jacob and Rudran 2003). However, the high cost and possible adverse effects of transmitters on the individual's behavior can be considered as possible disadvantages of this methodology (Jacob and Rudran 2003). In such context, the stable isotopes technique applied to trace the origin and movement of animals has been gaining strength in animal ecology (Rubenstein and Hobson 2004; Hobson and Wassenaar 2008).

The isotopic composition of animal tissues reflects the values of their local food chain and can be used to trace movements between isotopically distinct food webs (McKechnie 2004). Several biogeochemical processes can cause spatial variation in isotopic composition of food webs (Hobson 2008). However, the choice of animal tissue to be analyzed is a key part in the research design about the origin and movements of animals using stable isotopes, because different tissues reflect different temporal scales (Dalerum and Angerbjörn 2005). Metabolically inert tissues (e.g., nail, hair, and feather) reflect the isotopic composition of where they are synthesized, whereas metabolically active tissues (e.g., muscle, skin, and blood plasma) reflect the integration of dietary sources in different sites depending on their turnover rate (Bearhop et al. 2002; Ethier et al. 2010).

Hydrogen stable isotopes compositions (δD or δ^2H) are used in many studies of animal migration (Bowen et al. 2005), because the δD values have large amplitude (~500 ‰) and variation among distinct environments in nature (e.g., terrestrial and marine) (Wassenaar 2008). In addition, δD values vary according to the latitude, altitude, distance from the sea and precipitation (climatic process) (Dansgaard 1964; Chamberlain et al. 1997; Hobson and Wassenaar 1997; Hobson 2005; Hobson et al. 2012). Therefore, analysis of different parts of inert tissues may reveal the origin of migratory animals (Chamberlain et al. 1997). The application of hydrogen stable isotopes for this purpose has been particularly successful in studies with birds based on feather analysis (e.g., González-Prieto et al. 2011; Greenwood and Dawson 2011; Marquiss et al. 2012).

Hobson et al. (2003) provide an interesting example of the application of δD in the study of animal movement. The authors investigated the potential for this approach by measuring isotopic compositions ($\delta^{13}C$, δD, and $\delta^{15}N$) in tail feathers of eight species of hummingbirds along an altitudinal gradient (300–3,290 m) in the Andes Mountains of Ecuador. Avifauna inhabiting montane regions can move and feed between in isotopically distinct regions. This study found a strong relationship between $\delta^{13}C$, δD values of hummingbird feather, and elevation in the Ecuadorean Andes. In addition, the authors also discuss the possible origin of some species sampled in lower or higher elevation than their capture site.

Isotopic composition of other stable isotopes ($\delta^{13}C$, $\delta^{15}N$, and $\delta^{18}O$) individually or in combination can also be used to infer about origin or movements of organisms (Hobson 1999), for example, movements between environments with a predominance of C_3 and C_4 plants (Chisholm et al. 1986; Alisauskas et al. 1998); C_3 and crassulacean plants (Fleming et al. 1993) and marine and freshwater environments (Meyer-Rochow et al. 1992; Smith et al. 1996; Rosenblatt and Heithaus 2011).

Ogden et al. (2005) were able to quantify the proportional use of estuarine and terrestrial farmland resources by *Calidris alpina pacifica* (Dunlin) on the Fraser River Delta, British Columbia, using stable isotopes analysis ($\delta^{15}N$ and $\delta^{13}C$) of blood tissue. They found a great difference in intraspecific behavioral strategies, because the contribution of terrestrial farmland in diet ranging from 1 to 95 % between individuals. However, the proportion of diet attributed to terrestrial sources was 38 % when considering the mean isotopic values for Dunlin over four winters, 1997 through 2000. Juveniles showed higher terrestrial contribution to diet (43 %) than adults (35 %). Juveniles can forage more successfully in terrestrial farmland until gain experience to capture prey on the marine intertidal flats. In addition, Dunlin obtains most of its diet in these environments during periods of severe weather conditions. This study demonstrated that farmland terrestrial zone play an important role in the survival of Dunlin.

Maruyama et al. (2001) studied fluvial–lacustrine migrations of *Rhinogobius* sp. (landlocked goby: orange form) in the Lake Biwa water system, Japan, using stable isotope compositions (nitrogen and carbon isotope ratios). Previous reports showed that Lake Biwa has sediment and benthic animals with $\delta^{15}N$ values higher than those in the tributary rivers, then this isotope ratio could be used to trace fluvial–lacustrine migrations. This pattern of isotopic nitrogen distribution also occurred in *Rhinogobius* sp., and authors were able to detect that small individuals collected in the fluvial water body had spent their larval periods in the lake.

11.2.4 Tissue Turnover Rates

The application of stable isotopes analysis and correct interpretation of field data in wildlife studies rely on good estimates of tissue turnover rates. Isotopic turnover rate may be defined as the time that a tissue or whole consumer takes to reflect the isotopic composition of their diet (Tieszen et al. 1983; Gannes et al. 1998), in a process that occurs due to tissue growth and tissue replacement (MacAvoy et al. 2005).

The knowledge of differences in turnover rates is crucial to choose the appropriate tissue and to decide the sampling frequency in the individuals according to the objectives of a particular study, because the turnover rates varies between tissue types reflecting different timescales (Dalerum and Angerbjörn 2005; Rio and Carleton 2012). Tissues with a high turnover rate reflect the isotopic composition of food items consumed recently; on the other hand, tissues with low turnover rates reflect isotopic composition of food items consumed over a period of time (Hobson and Clark 1992, 1993).

The determination of turnover rates is also important for the interpretation of isotopic data, because accurate estimation of this parameter can improve interpretation of output isotope models (Phillips and Gregg 2001). Turnover rate may vary among individuals due to various factors, as growth rate, body size, and protein turnover rate (Newsome et al. 2010). Therefore, lab-controlled studies with the highest number possible of taxons are needed to better understand the factors that influence the dynamics of isotopic incorporation into animal tissues.

Experiments under controlled conditions have been conducted in order to determine the turnover rates of several tissues (e.g., Voigt et al. 2003; Seminoff et al. 2007; Murray and Wolf 2012; Storm-Suke et al. 2012). In this case, tissues of interest are analyzed to verify the time required for them to reflect the new consumer's diet. Murray and Wolf (2012) conducted studies with the desert tortoise (*Gopherus agassizii*) and observed a mean turnover rate of 126.7 days for red blood cells and 32.9 days for the plasma when analyzed $\delta^{13}C$. Hobson and Clark (1992) in controlled experiments with quails (*Coturnix japonica*) measured the carbon half-life of 11.4 days for the whole blood, 12.4 days for the muscle *pectoralis,* and 173.3 days for the bones collagen.

Oliveira (2003) investigating the dynamics of incorporation of carbon and nitrogen in the tissues of tambaqui fingerlings (*C. macropomum*) observed that the replacement rate of these elements vary according to the quality of the food source and the tissue functionality. In individuals fed with a diet based on C_3 plants turnover rate for the $\delta^{13}C$ and $\delta^{15}N$ was 42.7 and 28.9 d for liver, 77.9 and 85.5 d for muscle, and 104.5 and 125.7 d for scale, respectively. The turnover in visceral fat tested only for $\delta^{13}C$ was 184.7 d. In individuals with a diet based only on C_4 plants, liver reached equilibrium with the diet for $\delta^{13}C$ in 85.2 d. However, the author had observed that carbon substitution is faster than nitrogen substitution in all tissues in the C_4 plant-based diet.

Rosenblatt and Heithaus (2013) conducted an experiment under controlled conditions to estimate turnover rates for three tissues (scutes, red blood cells, and plasma) in American alligators (*Alligator mississippiensis*). This study tries to fill the gap in our understanding about turnover rates for crocodilians. Juvenile American alligators were housed in an enclosed and fed with equal amounts of food two times per week. Diet of pellet was changed to diet of channel catfish, and the tissues were collected over time. The isotope turnover rates of American alligators found in this study were considerably slower than those of most other taxa studied. The estimated $\delta^{13}C$ turnover rates for blood plasma, red blood cells, and scutes were 252, 566, and 590 d, respectively, and the estimated $\delta^{15}N$ turnover rates were 249.6, 1,109.2, and 414 d, respectively.

11.2.5 Ecotoxicology

Stable isotopes analysis is a powerful tool in ecotoxicological studies to understand the dynamics of contaminants on individuals and food webs (Crawford et al. 2008). This analysis provides a considerable advance to the ecotoxicology field by linking wild animals to their diet and contaminant source (Jardine et al. 2006).

Understanding the diet of organisms has a key role in ecotoxicological studies because most contaminants (heavy metals, organochlorine compounds, and other persistent contaminants) in animals are obtained by food consumption (Thomann and Connolly 1984; Hall et al. 1997). These contaminants pass through the process called bioaccumulation or biomagnification in the environment, in which concentrations of contaminants in consumers exceed those concentrations in diets (Gobas et al. 1993; Gobas and Morrison 2000). As the $\delta^{15}N$ also increases along the trophic chain as already seen, it is possible to relate it to the isotopic compositions values in food chains.

Jardine et al. (2006) considers three general categories of ecotoxicology studies that use stable isotopes analysis: qualitative linkages between dietary habits of animals and their contaminant concentrations, food web biomagnification studies, and quantitative assessments of habitat-specific foraging as a means of explaining biotic contaminant concentrations. In this context, the $\delta^{15}N$ enrichment in trophic webs helps to understand the contaminants paths along the food webs (Borga et al. 2004; Campbell et al. 2005), because there is a strong association between $\delta^{15}N$ enrichment and increasing concentrations of organochlorine and Hg contaminants (Broman et al. 1992; Kidd et al. 1995; Atwell et al. 1998; Campbell et al. 2005; Garcia and Carignan 2009). On the other hand, $\delta^{13}C$ values allow the traceability of foraging strategy and hence the specific sources of such contaminants (Crawford et al. 2008). There are several examples of studies adopting the approaches mentioned above (e.g., Atwell et al. 1998; Camusso et al. 1998; Thompson et al. 1998; Fox et al. 2002), however, we will address two papers in detail.

Di Beneditto et al. (2013) evaluate the trophic status and feeding ground of *Trichiurus lepturus* (ribbonfish) using total mercury concentration and stable isotope compositions ($\delta^{15}N$ and $\delta^{13}C$) during its ontogeny in the northern region of the State of Rio de Janeiro, south-eastern Brazil. Mercury is an environmental pollutant that bioaccumulates through the aquatic food chain and affects negatively human health. Mercury concentrations and $\delta^{15}N$ were different between sub-adult (planktivorous) and adult (carnivorous) specimens, indicating difference in trophic position of ontogenetic phases. However, the similarity of $\delta^{13}C$ values between sub-adults and adults suggest that both share the same feeding area (marine coastal waters). The mercury concentrations found in adults of *T. lepturus* are close to the tolerable limit for safe regular ingestion established by World Health Organization, so mercury levels in this fish species and environment should be monitored by public health authorities.

Das et al. (2004) studied trophic status, potential intraespecific segregation according to the source of prey and trace metals levels in harbor porpoise (*Phocoena phocoena relicta*) from the Black Sea. This environment has undergone an extensive human impact over the past decades which affected negatively wildlife populations. Harbor porpoises are at risk of disappearing and information on contaminant, their ecology and status are extremely important. The main result of this study was that differences in $\delta^{13}C$ between the sexes suggest that females use more the coastal environment (shallow waters) and males offshore habitats. The contaminant levels (hepatic Hg) in animals reflected the different exposure linked to coastal vs offshore feeding habitats.

11.3 What Next

Natural variations in abundance of stable isotopes provide an interesting tool for the study of energy flow systems. Currently, there is a growth in the use of stable isotopes analysis in animal ecology accompanying methodological development of the area (e.g. advances in analysis of stable isotope data and mass spectrometry). In this work, we addressed the main applications of isotopic analyses in wildlife studies emphasizing the ecological responses that can be achieved by this methodology. The topics were treated in a simple and concise form, and readers can deepen their knowledge in specific subjects in various articles and reviews available in the literature.

The stable isotope methodology has proven to be an interesting alternative in wildlife studies; however, some limitations need to be considered. Understanding the discrimination factors and routing processes in different tissues is needed to correct interpretation of isotopic data, beyond the knowledge of possible factors that may influence them (e.g., growth rate, age, and stress level). The call for controlled experiments to meet these goals has been performed by several authors (Gannes et al. 1997; Wolf et al. 2009) aiming to increase our ability to interpret values of stable isotopes.

Another limiting factor to be considered is the data resolution to distinguish different food sources and environments, for example, it is difficult to infer differences in diet contribution when food resources have similar isotopic composition. Technological developments in mass spectrometry, cost reduction, and concomitant analysis of a larger isotopes number can further improve the resolution studies. In this respect, technological development has enabled the use of compound-specific stable isotopes analysis of individual amino acids and fatty acids arouse great possibilities for studies in nutritional biochemistry of organisms.

The increased application of isotopic analysis in animal ecology also highlights the need to develop protocols for collecting and processing tissues. The method of tissues conservation, lipids extraction, and laboratory practices has significant effects on the isotopic compositions (Arrington and Winemiller 2002; Post et al. 2007).

Furthermore, there is a need to perform the isotopic monitoring trends over time. In the future, it is expected a major technological development and advances in the form of statistical analysis of isotopic data. The use of Bayesian inference in mixture models to estimate diet contribution incorporating uncertainty has provided more accurate estimates in recent years. The refinement of these types of analyzes might provide a better interpretation of isotopic patterns.

References

Adams TS, Sterner RW (2000) The effect of dietary nitrogen on trophic level ^{15}N enrichment. Limnol Oceanogr 45:601–607

Alisauskas RT, Klaas EE, Hobson KA, Ankney CD (1998) Stable-carbon isotopes support use of adventitious color to discern winter origins of lesser snow geese. J Field Ornithol 69:262–268

Ambrose SH (1990) Preparation and characterization of bone and tooth collagen for isotopic analysis. J Archaeol Sci 17:431–451

Arrington DA, Winemiller KO (2002) Preservation effects on stable isotope analysis of fish muscle. Trans Am Fisher Soc 131:337–342

Atwell L, Hobson KA, Welch HE (1998) Biomagnification and bioaccumulation of mercury in an arctic marine food web: insights from stable nitrogen isotope analysis. Can J Fish Aquat Sci 55:1114–1121

Bearhop S, Waldron S, Votier SC, Furness RW (2002) Factor that influence assimilation rates and fractionation of nitrogen and carbon stable isotopes in avian blood and feathers. Physiol Biochem Zool 75:451–458

Boecklen WJ, Yarnes CT, Cook BA, James AC (2011) On the use of stable isotopes in trophic ecology. Annu Rev Ecol Evol Syst 42:411–440

Borga K, Fisk AT, Hoekstra PF, Muir DCG (2004) Biological and chemical factors of importance in the bioaccumulation and trophic transfer of persistent organochlorine contaminants in Arctic marine food webs. Environ Toxicol Chem 23:2367–2385

Borteiro C, Gutiérrez F, Tedrosa M, Kolenc F (2009) Food habits of the Broad-snouted Caiman (*Caiman latirostris*: Crocodylia, Alligatoridae) in northwestern Uruguay. Stud Neoptrop Fauna Envir 44:31–36

Bowen GJ, Wassenaar LI, Hobson KA (2005) Global application of stable hydrogen and oxygen isotopes to wildlife forensics. Oecologia 143:337–348

Broman D, Naf C, Rolff C, Zebuhr Y, Fry B, Hobbie J (1992) Using ratios of stable nitrogen isotopes to estimate bioaccumulation and flux of polychlorinated dibenzo-p-dioxins (PCDDs) and dibenzofurans (PCDFs) in two food chains from the northern Baltic. Environ Toxicol Chem 11:331–345

Campbell LM, Fisk AT, Wang X, Kock G, Muir DCG (2005) Evidence for biomagnification of rubidium in freshwater and marine foodwebs. Can J Fish Aquat Sci 62:1161–1167

Camusso M, Martinotti W, Balestrini R, Guzzi L (1998) C and N stable isotopes and trace metals in selected organisms from the River Po Delta. Chemosphere 37:2911–2920

Caut S, Angulo E, Courchamp F (2009) Variation in discrimination factors ($\Delta^{15}N$ and $\Delta^{13}C$): the effect of diet isotopic values and applications for diet reconstruction. J Appl Ecol 46:443–453

Chamberlain CP, Blum JD, Holmes RT, Feng XH, Sherry TW, Graves GR (1997) The use of isotope tracers for identifying populations of migratory birds. Oecologia 109:132–141

Chisholm B, Driver J, Dube S, Schwarcz HP (1986) Assessment of prehistoric bison foraging and movement patterns via stable-carbon isotope analysis. Plains Anthropol 31:193–205

Clementz MT (2012) New insight from old bones: stable isotope analysis of fossil mammals. J Mammal 93:368–380

Crawford K, McDonald RA, Bearhop S (2008) Applications of stable isotope techniques to the ecology of mammals. Mammal Rev 38:87–107

Dalerum F, Angerbjörn A (2005) Resolving temporal variation in vertebrate diets using naturally occurring stable isotopes. Oecologia 144:647–658

Dansgaard W (1964) Stable isotopes in precipitation. Tellus 5:436–468

Das K, Holsbeek L, Browning J, Siebert U, Birkun A, Bouquegneau JM (2004) Trace metal and stable isotope measurements ($\delta^{13}C$ and $\delta^{15}N$) in the harbour porpoise *Phocoena phocoena relicta* from the Black sea. Environ Pollut 131:197–204

DeNiro MJ (1985) Postmortem preservation and alteration of in vivo bone collagen isotope ratios in relation to palaeodietary reconstruction. Nature 317:806–809

DeNiro MJ (1987) Stable isotopy and archaeology. Am Sci 75:182–191

DeNiro MJ, Epstein S (1978) Influence of diet on the distribution of carbon isotopes in animals. Geochim Cosmochim Acta 42:495–506

DeNiro MJ, Epstein S (1981) Influence of diet on the distribution of nitrogen isotopes in animals. Geochim Cosmochim Acta 45:341–351

Di Beneditto APM, Bittar VT, Rezende CE, Camargo PB, Kehrig HÁ (2013) Mercury and stable isotopes ($\delta^{15}N$ and $\delta^{13}C$) as tracers during the ontogeny of *Trichiurus lepturus*. Neotrop Ichthyol 11:211–216

Erhardt EB, Bedrick EJ (2013) A Bayesian framework for stable isotope mixing models. Environ Ecol Stat 20:377–397

Estrada JA, Rice AN, Lutcavage ME, Skomal GB (2003) Predicting trophic position in sharks of the north-west Atlantic ocean using stable isotope analysis. J Mar Biol Ass UK 83:1347–1350

Ethier DM, Kyle CJ, Kyser TK, Nocera JJ (2010) Variability in the growth patterns of the cornified claw sheath among vertebrates: implications for using biogeochemistry to study animal movement. Can J Zool 88:1043–1051

Fleming TH, Nuñez RA, Sternberg LSL (1993) Seasonal changes in the diets of migrant and non-migrant nectarivorous bats as revealed by carbon stable isotope analysis. Oecologia 94:72–75

Fox GA, Grasman KA, Hobson KA, Williams K, Jeffrey D, Hanbridge B (2002) Contaminant residues in tissues of adult and prefledged herring gulls from the Great Lakes in relation to diet in the early 1990s. J Great Lakes Res 28:643–663

Fry B (1991) Stable isotope diagrams of freshwater food webs. Ecology 72:2293–2297

Fry B (2006) Stable isotope ecology. Springer, Berlin

Gannes LZ, O'Brien D, del Rio CM (1997) Stable isotopes in animal ecology: assumptions, caveats, and a call for laboratory experiments. Ecology 78:1271–1276

Gannes LZ, Rio CM, Kouch P (1998) Natural abundance variations in stable isotopes and their potential uses in animal physiological ecology. Comp Biochem Physiol 119:725–737

Garcia E, Carignan R (2009) Mercury concentrations in fish from forest harvesting and fire-impacted Canadian Boreal lakes compared using stable isotopes of nitrogen. Environ Toxicol Chem 24:685–693

Gobas FAPC, Zhang X, Wells R (1993) Gastrointestinal magnifications: the mechanism of biomagnification and food chain accumulation of organic chemicals. Environ Sci Technol 27:2855–2863

Gobas FAPC, Morrison HA (2000) Bioconcentration and bioaccumulation in the aquatic environment. In: Boethling R, Mackay D (eds) Handbook of property estimation methods for chemicals: environmental and health sciences. CRC Press LLC, Boca Raton, pp 189–231

González-Prieto AM, Hobson KA, Bayly NJ, Gómez C (2011) Geographic origins and timing of fall migration of the veery in Northern Colombia. Condor 113:860–868

Greenwood JL, Dawson RD (2011) Correlates of deuterium (δD) enrichment in the feathers of adult American Kestrels of known origin. Condor 113:555–564

Hall BD, Bodaly RA, Fudge RJP, Rudd JWM, Rosenberg DM (1997) Food as the dominant pathway of methylmercury uptake by fish. Water Air Soil Pollut 100:13–24

Hobson KA (1999) Tracing origins and migration of wildlife using stable isotopes: a review. Oecologia 120:314–326

Hobson KA (2005) Using stable isotopes to trace long-distance dispersal in birds and other taxa. Divers Distrib 11:157–164

Hobson KA (2008) Using endogenous and exogenous markers in bird conservation. Bird Conserv Internat 18:S174–S199

Hobson KA, Clark RW (1992) Assessing avian diets using stable isotopes: turnover of carbon-13. Condor 94:181–188

Hobson KA, Clark RG (1993) Turnover of $\delta^{13}C$ in cellular and plasma fractions of blood: implications for nondestructive sampling in avian dietary studies. Auk 110:638–641

Hobson KA, Wassenaar LI (1997) Linking brooding and wintering grounds of neotropical migrant songbirds using stable hydrogen isotopic analysis of feathers. Oecologia 109:142–148

Hobson KA, Wassenaar LI (2008) Tracking animal migration with stable isotopes. Academic Press, San Diego

Hobson KA, Alisaiskas RT, Clark RG (1993) Stable-nitrogen isotope enrichment in avian tissues due to fasting and nutritional stress: implications for isotopic analysis of diet. Condor 95:388–394

Hobson KA, Van Wilgenburg SL, Wassenaar LI, Larson K (2012) Linking hydrogen (δ^2H) isotopes in feathers and precipitation: sources of variance and consequences for assignment to isoscapes. PLoS ONE 7:e35137

Hobson KA, Wassenaar LI, Milá B, Lovette I, Dingle C, Smith TB (2003) Stable isotopes as indicators of altitudinal distributions and movements in an Ecuadorean hummingbird community. Oecologia 136:302–308

Inguer R, Bearhop S (2008) Applications of stable isotope analyses to avian ecology. Ibis 150:447–461

Jacob AA, Rudran R (2003) Radiotelemetria em estudos populacionais. In: Cullen Júnior L, Rudan R, Valladares-Padua C (eds.) Métodos de estudos em biologia da conservação & manejo da vida silvestre. Editora da Universidade Federal do Paraná, Curitiba, Brasil, 285–342

Jardine TD, Kidd KA, Fisk AT (2006) Applications, considerations, and sources of uncertainty when using stable isotope analysis in ecotoxicology. Environ Sci Technol 40(24):7501–7511

Kelly JF (2000) Stable isotopes of carbon and nitrogen in the study of avian and mammalian trophic ecology. Can J Zool 78:1–27

Kidd KA, Schindler DW, Hesslein RH, Muir DCG, Lockhart WL, Hesslein RH (1995) High concentrations of toxaphene in fishes from a subarctic lake. Science 269:240–242

Koch PL (2007) Isotopic study of the biology of modern and fossil vertebrates. In: Michener RH, Lajtha K (eds) Stable isotopes in ecology and environmental science. Blackwell, Boston, MA, pp 99–154

Lara NRF, Marques TS, Montelo KM, Ataídes AG, Verdade LM, Malvásio A, Camargo PB (2012) A trophic study of the sympatric Amazonian freshwater turtles *Podocnemis unifilis* and *Podocnemis expansa* (Testudines, Podocnemidae) using carbon and nitrogen stable isotope analyses. Can J Zool 90:1394–1401

Layman CA, Araújo MS, Boucek R, Harrison E, Jud ZR, Matich P, Hammerschlag-Peyer CM, Rosenblatt A, Vaudo J, Yeager LA, Post DM, Bearhop S (2012) Applying stable isotopes to examine food web structure: an overview of analytical tools. Biol Rev 87:545–562

Litvaitis JA (2000) Investigating food habits of terrestrial vertebrates. In: Boitani L, Fuller TK (eds) Research techniques in animal ecology: controversies and consequences. Columbia University Press, New York, pp 165–190

MacAvoy SE, Macko SA, Arneson LS (2005) Growth versus metabolic tissue replacement in mouse tissues determined by stable carbon and nitrogen isotope analysis. Can J Zool 83:631–641

MacFadden BJ (2000) Cenozoic mammalian herbivores from the Americas: reconstructing ancient diets and terrestrial communities. Annu Rev Ecol Syst 31:33–59

MacFadden B, Solounias N, Cerling TE (1999) Ancient diets, ecology, and extinction of 5-million-year-old horses from Florida. Science 283:824–827

Macko SA, Fogel-Estep ML, Engel MH, Hare PH (1986) Kinetic fractionation of nitrogen isotopes during amino acid transformation. Geochim Cosmochim Acta 50:2143–2146

Macko SA, Fogel-Estep ML, Engel MH, Hare PH (1987) Isotopic fractionation of nitrogen and carbon in the synthesis of amino acids by microorganisms. Chem Geol 65:79–92

Manetta GI, Benedito-Cecilio E, Martinelli M (2003) Carbon source and trophic position of the main species of fishes of Baía river, Paraná river floodplain, Brazil. Braz J Biol 63:283–290

Marques TS, Lara NRF, Bassetti LAB, Piña CI, Camargo PB, Verdade LM (2013) Intraspecific isotopic niche variation in broad-snouted caiman (*Caiman latirostris*). Isot Environ Health Stud 49(3):325–335

Marquiss M, Newton I, Hobson KA, Kolbeinsson Y (2012) Origins of irruptive migrations by common crossbills *Loxia curvirostra* into northwestern Europe revealed by stable isotope analysis. Ibis 154:400–409

Martinelli LA, Ometto JPHB, Ferraz ES, Victoria RL, Camargo PB, Moreira MZ (2009) Desvendando Questões ambientais com Isótopos estáveis. Oficina de Textos, São Paulo

Maruyama A, Yamada Y, Yuma M, Rusuwa B (2001) Stable nitrogen and carbon isotope ratios as migration tracers of a landlocked goby, *Rhinogobius* sp. (the orange form), in the Lake Biwa water system. Ecol Res 16:697–703

McCutchan JH Jr, Lewis WM Jr, Kendall C, McGrath CC (2003) Variation in trophic shift for stable isotope ratios of carbon, nitrogen, and sulfur. Oikos 102:378–390

McKechnie AE (2004) Stable isotopes: powerful new tools for animal ecologists. S Afr J Sci 100:131–134

Melo MTQ (2002) Dieta do *Caiman latirostris* no sul do Brasil. In: Verdade LM, Larriera A (eds.) Conservação e manejo jacarés e crocodilos da América Latina. C.N. Editora, Piracicaba, 116–125

Meyer-Rochow VB, Cook I, Hendy CH (1992) How to obtain clues from the otoliths of an adult fish about the aquatic environment it has been in as a larva. Comp Biochem Physiol 103A:333–335

Millspaugh JJ, Marzluff JM (2001) Radio tracking and animal populations. Academic Press, San Diego

Minagawa M, Wada E (1984) Stepwise enrichment of ^{15}N along food chains: further evidence and the relationship between ^{15}N and animal age. Geochim Cosmochim Acta 48:1135–1140

Murray IW, Wolf BO (2012) Carbon incorporation rates and diet-to-tissue discrimination in ectotherms: tortoises are really slow. Physiol Biochem Zool 85:96–105

Newsome SD, Clementz MT, Koch PL (2010) Using stable isotope biogeochemistry to study marine mammal ecology. Mar Mammal Sci 26:509–572

Ogden LJE, Hobson KA, Lank DB, Bittman S (2005) Stable isotope analysis reveals that agricultural habitat provides an important dietary component for nonbreeding Dunlin. Avian Conserv Ecol 1:3

Olive PJW, Pinnegar JK, Polunin NVC, Richards G, Welch R (2003) Isotope trophic-step fractionation: a dynamic equilibrium model. J Anim Ecol 72:608–617

Oliveira ACB (2003) Isótopos estáveis de C e N como indicadores qualitativo e quantitativo da dieta do tambaqui (*Colossoma macropomum*) da Amazônia Central. Thesis in Sciences of the Centro de Energia Nuclear na Agricultura, Universidade de São Paulo, Piracicaba, Brazil

Oliveira ACB (2006) Seasonality of energy sources of *Colossoma macropomum* in a floodplain lake in the Amazon—lake Camaleão, Amazonas, Brazil. Fish Manag Ecol 13:135–142

Parkington J (1991) Approaches to dietary reconstruction in the Western Cape: are you what you have eaten? J Archaeol Sci 18:331–342

Parnell AC, Inger R, Bearhop S, Jackson AL (2010) Source partitioning using stable isotopes: coping with too much variation. PLoS ONE 5:e9672

Parnell AC, Phillips DL, Bearhop S, Semmens BX, Ward EJ, Moore JW, Jackson AL, Grey J, Kelly DJ, Inger R (2013) Bayesian stable isotope mixing models. http://arxiv.org/pdf/1209.6457.pdf. Accessed 20 Sept 2013

Pate FD (1997) Bone chemistry and paleodiet: reconstructing prehistoric subsistence-settlement systems in Australia. J Anthropol Archaeol 16:103–120

Phillips DL, Gregg JW (2001) Uncertainty in source partitioning using stable isotopes. Oecologia 127:171–179

Phillips DL, Gregg JW (2003) Source partitioning using stable isotopes: coping with too many sources. Oecologia 136:261–269

Post DM (2002) Using stable isotopes to estimate trophic position: models, methods, and assumptions. Ecology 83:703–718

Post DM, Layman CA, Arrington DA, Takimoto G, Quattrochi J, Montana CG (2007) Getting to the fat of the matter: models, methods and assumptions for dealing with lipids in stable isotope analyses. Oecologia 152:179–189

Pritchard PCH, Trebbau P (1984) The turtles of Venezuela. Society for the Study of Amphibians and Reptiles, Athens, Ohio

Ramos R, González-Solís J (2012) Trace me if you can: the use of intrinsic biogeochemical markers in marine top predators. Front Ecol Environ 10:258–266

Rio CMD, Carleton S (2012) How fast and how faithful: the dynamics of isotopic incorporation into animal tissues. J Mammal 93:353–359

Rosenblatt AE, Heithaus MR (2011) Does variation in movement tactics and trophic interactions among American alligators create habitat linkages? J Anim Ecol 80:786–798

Rosenblatt AE, Heithaus MR (2013) Slow isotope turnover rates and low discrimination values in the American alligator: implications for interpretation of ectotherm stable isotope data. Physiol Biochem Zool 86:137–148

Roth JD, Hobson KA (2000) Stable carbon and nitrogen isotopic fractionation between diet and tissue of captive red fox: implications for dietary reconstruction. Can J Zool 78:848–852

Rubenstein DR, Hobson K (2004) From birds to butterflies: animal movement patterns and stable isotopes. Trend Ecol Evol 19:256–263

Seminoff JA, Bjorndal KA, Bolten AB (2007) Stable carbon and nitrogen isotope discrimination and turnover in pond sliders *Trachemys scripta*: insights for trophic study of freshwater turtles. Copeia 2007:534–542

Smith RJ, Hobson KA, Koopman HN, Lavigne DM (1996) Distinguishing between populations of fresh and saltwater harbor seals (*Phoca vitulina*) using stable-isotope ratios and fatty acid profiles. Can J Fish Aquat Sci 53:272–279

Storm-Suke A, Norris DR, Wassenaar LI, Chin E, Nol E (2012) Factors influencing the turnover and net isotopic discrimination of hydrogen isotopes in Proteinaceous tissue: experimental results using Japanese Quail. Physiol Biochem Zool 85:376–384

Thomann RV, Connolly JP (1984) Model of PCB in the Lake Michigan lake trout food chain. Environ Sci Technol 18:65–71

Thompson DR, Bearhop S, Speakman JR, Furness RW (1998) Feathers as a means of monitoring mercury in seabirds: insights from stable isotope analysis. Environ Pollut 101:193–200

Tieszen LL, Boutton TW, Tesdahl KG, Slade NA (1983) Fractionation and turnover of stable carbon isotopes in animal tissues: implication for $\delta^{13}C$ analysis of diet. Oecologia 57:32–37

Vander Zanden MJ, Cabana G, Rasmussen JB (1997) Comparing trophic position of freshwater fish calculated using stable nitrogen isotope ratios ($\delta^{15}N$) and literature dietary data. Can J Fish Aquat Sci 54:1142–1158

Van Klinken GJ (1999) Bone collagen quality indicators for paleodietary and radiocarbon measurements. J Archaeol Sci 26:687–695

Voigt CC, Matt F, Michener R, Kunz TH (2003) Low turnover rates of carbon isotopes in tissues of two nectar-feeding bat species. J Exp Biol 206:1419–1427

Wassenaar LI (2008) An introduction to light stable isotopes for use in terrestrial animal migration studies. In: Hobson K, Wassenaar LI (eds) Tracking animal migration with stable isotopes. Academic Press, San Diego, pp 21–44

Wolf N, Carleton SA, del Rio CM (2009) Ten years of experimental animal isotopic ecology. Funct Ecol 23:17–26

Part III
Governance

Chapter 12
Multi-taxa Surveys: Integrating Ecosystem Processes and User Demands

William E. Magnusson, Ben Lawson, Fabricio Baccaro, Carolina Volkmer de Castilho, J. Guy Castley, Flavia Costa, Debora P. Drucker, Elizabeth Franklin, Albertina P. Lima, Regina Luizão, Fernando Mendonça, Flávia Pezzini, Juliana Schietti, José Julio Toledo, Ana Tourinho, Luciano M. Verdade and Jean-Marc Hero

Abstract Globally, natural resource management agencies are increasingly recognizing the importance of long-term ecological research (LTER) for monitoring biodiversity, ranging from relatively simple, known, local-level issues, such as managing tourist impacts in a conservation park, to more complex, multifaceted, pervasive, and far-reaching impacts, such as global climate change. Much previous literature has confused protocols for LTER projects to answer current research questions, with developing a system for long-term ecological monitoring. Contrary to perceptions that these LTER systems are not driven by well-defined objectives, we argue that LTER systems can be designed and implemented with the specific objective of providing a basis for both LTER projects and long-term monitoring. We present an overview of RAPELD,

W. E. Magnusson (✉) · F. Baccaro · F. Costa · E. Franklin · A. P. Lima · R. Luizão · F. Pezzini · J. Schietti · J. J. Toledo · A. Tourinho
Instituto Nacional de Pesquisas da Amazonia (INPA), Caixa Postal 478, Manaus, AM 69011-970, Brazil
e-mail: bill@inpa.gov.br

F. Baccaro
e-mail: fabricera@gmail.com

F. Costa
e-mail: flaviacosta001@gmail.com

E. Franklin
e-mail: beth@inpa.gov.br

A. P. Lima
e-mail: lima@inpa.gov.br

R. Luizão
e-mail: rccl@inpa.gov.br

F. Pezzini
e-mail: flaviapezzini@gmail.com

J. Schietti
e-mail: jujuschietti@gmail.com

L. M. Verdade et al. (eds.), *Applied Ecology and Human Dimensions in Biological Conservation*, DOI: 10.1007/978-3-642-54751-5_12,

an LTER system developed in Brazil, with comparable infrastructure established in Australia and Nepal. The standardized biodiversity infrastructure and research platform provides a long-term basis for powerful multi-disciplinary, multi-scale analyses.

12.1 Introduction

In 2002, the US Secretary of Defense, Donald Rumsfeld, used the phrase "unknown unknowns" to describe critically important aspects of military defense that are essentially unpredictable until they start to have an effect. Taleb (2007) more generally used the term "black swans" to describe rare and unpredictable phenomena that have a disproportionate effect on human lives, but focused on economic aspects. In the context of biodiversity conservation and management, such black swans include unpredictable long-term global issues, such as climate change (Doak et al. 2008; Wintle et al. 2010).

J. J. Toledo
e-mail: jjuliotoledo@gmail.com

A. Tourinho
e-mail: amtourinho@gmail.com

B. Lawson · J. Guy Castley · J.-M. Hero
Environmental Futures Centre, School of Environment, Griffith University, Gold Coast Campus, Gold Coast, QLD 4222, Australia
e-mail: zzbenlaw@uqconnect.net

J. Guy Castley
e-mail: g.castley@griffith.edu.au

J.-M Hero
e-mail: m.hero@griffith.edu.au

C. V. de Castilho
Centro de Pesquisa Agroflorestal de Roraima, Empresa Brasileira de Pesquisa Agropecuária (EMBRAPA), BR 174, Km 8, Distrito Industrial, Boa Vista, RR 69301-970, Brazil
e-mail: carolina@cpafrr.embrapa.br

D. P. Drucker
EMBRAPA Monitoramento por Satélite, Av. Soldado Passarinho, 303, Campinas, SP 13070-115, Brazil
e-mail: debora@cnpm.embrapa.br

F. Mendonça
Instituto de Saúde e Biotecnologia—ISB, Universidade Federal do Amazonas—UFAM, Campus Universitário do Médio Solimões, Estrada Coari-Mamiá, 305, Coari, AM 69460-000, Brazil
e-mail: fernandopmendonca@gmail.com

L. M. Verdade
Centro de Energia Nuclear na Agricultura, Universidade de São Paulo, Caixa Postal 96, Piracicaba, SP 13416-000, Brazil
e-mail: lmverdade@usp.br

The black swan concept goes to the heart of a long-running debate about the value of long-term ecological monitoring and whether the approaches advocated by various authors are able to achieve their stated objectives (Magnusson et al. 2005, 2008; Ferraz et al. 2008; Lindenmayer and Likens 2009; Haughland et al. 2010). Much of this debate we feel can be attributed to a misunderstanding of the fundamental differences between designing a long-term ecological research (LTER) project focused on particular questions (Lindenmayer and Likens 2009), and designing an LTER system for addressing current management questions, but which can also detect "black swans" over much greater timescales (Magnusson et al. 2005; Haughland et al. 2010; Costa and Magnusson 2010; Hero et al. 2010).

Designing a monitoring program can benefit from the attention paid to a limited range of management goals, resulting in narrowly focused scientific research projects. This approach works well if all stakeholders are convinced of the relevance of those goals, and it is reasonable to assume that other stakeholders, or goals, will not be forced onto the study. However, these assumptions are likely to apply only to geographically and temporally limited studies.

The LTER project approach to scientific analysis is well suited to monitoring, which has the objective of estimating parameters for pre-established models. However, it is increasingly becoming evident that, without modification, this approach cannot prepare us for early detection of black swans, which are generally unpredicted because our models, rather than the parameter estimates, are badly specified.

Firstly, the cost of implementing an optimal experimental design for each and every known environmental threat within a particular scientific researcher's universe of interest is beyond any reasonable expectation for financing within the foreseeable future (Field et al. 2005). Secondly, even if it were possible to finance such complex experimental designs, these studies would not self-organize into a system that would optimize our chances of detecting and dealing with black swans (Wintle et al. 2010). We argue that developing LTER systems to effectively understand and manage biodiversity requires a paradigm shift in LTER approaches.

In this chapter, we present some outcomes from the RAPELD system for use in LTER networks—a system designed to answer specific research questions in a long-term monitoring framework. It was designed for detecting long-term trends in biodiversity across longitudinal and latitudinal gradients at a global scale and to maximize the chances of detecting black swans. It does this by implementing infrastructure that is useful to a wide range of stakeholders with different objectives and scales of interest, while maintaining the flexibility to deal with specific threats and evolving research questions.

To be of greatest use to the widest range of stakeholders, the RAPELD LTER system was designed around the following eight fundamental requirements:

1. Be technically and spatially standardized.
2. Permit standardized surveys of all taxa.
3. Be large enough to permit survey of all taxa and ecosystem processes.
4. Be modular to permit surveys over large areas.

5. Be compatible with preexisting initiatives.
6. Can be implemented with available human and monetary resources.
7. Provide data to stakeholders in a reasonable time.
8. Be compatible and integrated with large-scale remote sensing capabilities.

Where possible, surveys should result in abundance estimates, or at least permit evaluation of detectability biases (see Chaps. 9 and 10). Details behind the logic and descriptions of implementation for each of these can be found in Costa and Magnusson (2010), Hero et al. (2010), and the PPBio Web site (http://ppbio.inpa.gov.br). However, even without the details, it should be clear why principles such as these are necessary to attend to the demands of the wide range of biodiversity stakeholders, and a number of these points are raised by others committed to establishing monitoring networks (Westoby 1991; Parr et al. 2002; Buckland et al. 2005; Abbott and Le Maitre 2010; Gardner 2010). Stakeholders contribute to the system in a variety of ways, including providing manpower, maintaining field infrastructure, developing remote sensing techniques, providing access to their lands, identifying specimens, and undertaking laboratory analyses. It is their system.

Biodiversity monitoring represents an extreme case in which the stakeholders, objectives, and funding opportunities are spatially and temporally scale dependent, and there is no overarching institutional framework to coordinate monitoring efforts. We will illustrate this with examples from the Brazilian Program for Biodiversity Research (PPBio), and the use of RAPELD (Magnusson et al. 2005) methodology. We do not purport to describe the myriad of ecological interactions in global ecosystems but aim instead to show how the RAPELD system allows different stakeholders to make use of the same infrastructure, despite different and sometimes changing objectives.

12.2 How to Catch a Black Swan

Here, we present some examples of the use of RAPELD infrastructure that provide evidence that LTER systems are much better equipped for detecting black swans. These examples are not exhaustive, but they demonstrate the ability to detect black swans using a well-designed LTER system that provides infrastructure and a framework for answering unexpected multi-scale and multi-disciplinary ecological questions.

1. Reserva Ducke is one of the most accessible and intensively studied field sites in the Neotropics (Pitman et al. 2011), and millions of dollars had been spent in trying to document the biodiversity in the reserve (e.g., Adis 2002; Ribeiro et al. 1999). However, many new species for Reserva Ducke were discovered when scientists were provided with infrastructure (permanent trails and plots) to systematically survey the entire reserve, including birds (Cintra 2008), plants (Costa et al. 2008, 2009), fish (Mendonça et al. 2008), and frogs (Menin et al. 2008a, b). One of the most unexpected results was that *Atelopus spumarius*, which belongs to one of the most extinction-prone genera of frogs,

occurs only in the western drainage of the reserve where the water is slightly less acid (Menin et al. 2008a). That important species had not been detected in the reserve despite many long-term studies undertaken before the field infrastructure was established (e.g., Magnusson et al. 1999; Hero et al. 2001).

Earlier studies, although carried out by experts, had failed to detect many relatively large vascular plants and vertebrates of Reserva Ducke, and these results were replicated in other areas. A graduate student, in a survey of a standard RAPELD grid in the area of influence of the Balbina hydroelectric dam, encountered 58 species of frogs, whereas much more extensive studies by specialists in a much greater area, including more obviously distinct habitats, had revealed only 48 species (Condrati 2009). RAPELD methodology, using relatively inexperienced observers, encountered about twice as many *species of amphibians* for the same field effort as directed surveys by specialists elaborating environmental impact statements (Goralewski 2008). Apparently, the advantages of systematic sampling far outweigh the advantages of using experts without the benefit of standardized field infrastructure.

2. The existence of long-term ecological research sites using shared infrastructure allows multi-disciplinary studies that were not expected or designed and allow better interpretation of short-term studies using the same methodology. For instance, Dias et al. (2010) used data from an LTER site that also used RAPELD methodology to evaluate the probable effect of seasonality revealed by a short-term study of effects of logging on fish.
3. Remote Sensing: Using standard infrastructure systematically distributed across the landscape facilitates integrating remote sensing with ground truthing. The possibilities for feedbacks between remote sensing and biodiversity surveys can be illustrated with two examples from Amazonian RAPELD sites. Use of Shuttle Radar Topography Mission (SRTM) images is now one of the most useful and robust methods to produce topographic maps over large areas. However, SRTM reflects off forest canopies and may be affected by surface water, and it was not clear how accurate data derived from SRTM was for estimation of altitude over small areas. Schietti et al. (2007) used data from several RAPELD LTER sites to calibrate SRTM data revealing altitudinal precision errors under the canopy over scales of tens of km ($r^2 = 0.7$) with further bias over waterlogged ground. The ability to undertake these calibration measures was fortuitous as the elevation data collected from the LTER plots were not measured with this objective in mind. Nonetheless, these data are proving useful to both remote sensing modelers and biodiversity specialists.

Light detection and ranging (LIDAR) methods use point height data to produce surface topographical strata and can be used to generate structural images of forest canopies. It may therefore be possible to calibrate LIDAR data to estimate arboreal biomass (Lefsky et al. 2002). This has become the objective of a consortium of Brazilian and North American researchers (http://www.amazonpire.org/) who are using data from RAPELD and other long-term monitoring sites in Amazonia to calibrate airborne LIDAR. These data are also being calibrated against ground-based LIDAR and in the future will be used to calibrate satellite LIDAR (M. A. Lefsky, pers. comm.). This is another example preparing for a black swan, in which LIDAR technology was not envisioned when the RAPELD grids were being installed, though the calibration would not have been as effective if remote sensing in general had not been taken into account in designing the RAPELD system (Costa and Magnusson 2010).

4. Many studies have shown biomass accumulation in Amazonian forests. However, only studies undertaken using RAPELD infrastructure have been able to show within-site variation in biomass accumulation due to soils, and short-term (2-year) variation in soil-biomass accumulation relationships (Castilho et al. 2006, 2010). This biomass comparison can be linked to climate change, and data are now comparable with other RAPELD LTER sites nationally and internationally. Although the total biomass per hectare is much greater in the Amazonian rainforest site, the proportion of biomass in each 10-cm-diameter-at-breast-height size class is very similar in Australian eucalypt forest (Fig. 12.1). The proportion of biomass in each size at Reserva Ducke (Castilho et al. 2006) predicts 95 % of the variation between size classes in Karawatha measured by Butler (2007), a result totally unexpected before the comparison was made.

12.2.1 Examples of LTER Projects that Used RAPELD Infrastructure

RAPELD LTER infrastructure has proven useful for studies of carbon stocks (Castilho et al. 2006, 2010), decomposition processes (Braga-Neto et al. 2008; Toledo et al. 2009, 2011, 2012), and distributions of plants (Costa et al. 2005, 2009; Kinupp and Magnusson 2005; Drucker et al. 2008), fish (Mendonça et al. 2005; Pazin et al. 2006; Espírito-Santo et al. 2009), amphibians (Menin et al. 2007, 2008b), mammals (Mendes Pontes et al. 2008, 2012; Calzada et al. 2008), and invertebrates (Oliveira et al. 2009; Rodrigues et al. 2010). It has revealed gaps in our knowledge with regard to the effects of forestry (Castilho et al. 2006; Dias et al. 2010) and legislation to protect riparian areas (Drucker et al. 2008; Bueno et al. 2012). It is allowing integration of LTER and ILTER sites in ways that were not previously possible. This is not because RAPELD infrastructure is necessarily optimal for surveys of any particular taxon or ecosystem process. Its strength comes from planning which encourages interactions among researchers from diverse fields as well as with management and industry partners. It would appear that the usefulness of the infrastructure provided by the RAPELD system is more limited by our imagination than by any limits in the design.

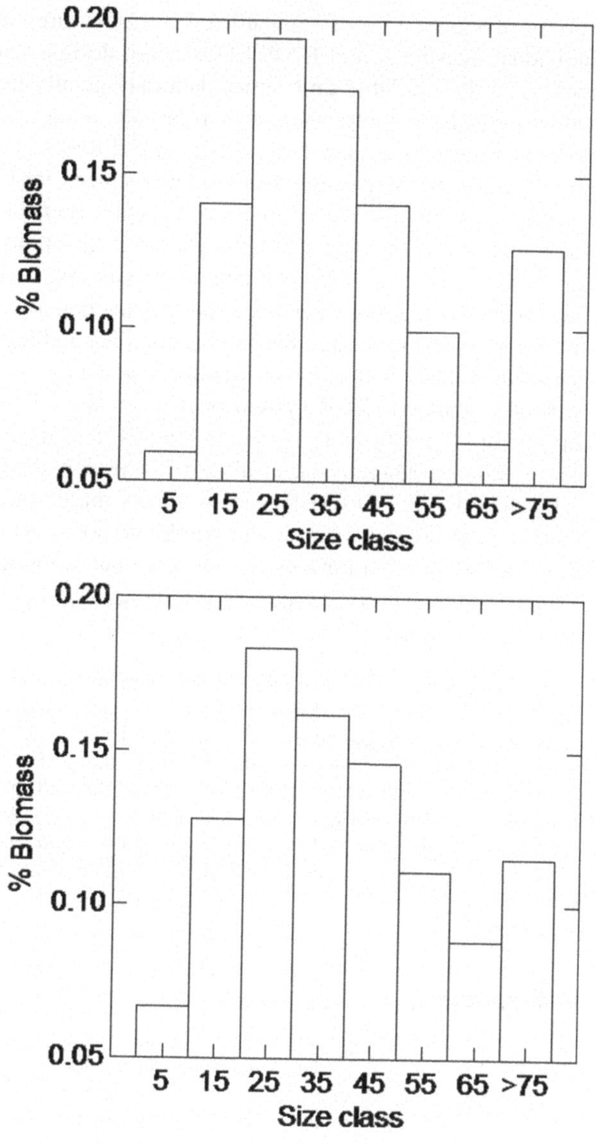

Fig. 12.1 Proportion of biomass in 10-cm-diameter-at-breast-height size classes of trees in Reserva Ducke, Amazonas, Brazil, and Karawatha Reserve, Queensland, Australia

12.3 Conclusion

The RAPELD LTER system is unashamedly standardized. It is also unashamedly flexible. No system can guarantee answers to all questions on all scales. However, systems should be designed to be used for a wide range of questions across a great variety of scales. Use of standardized infrastructure cannot replace thinking about individual questions, and RAPELD was not designed for blind application, though looking at the available time series data may greatly facilitate planning for an individual study. New approaches, such as NEON in the USA (http://www.neoninc.org/), ATBI in Europe (Eymann et al. 2010), and TERN-ACEAS in Australia (http://www.tern.org.au/), are increasingly recognizing the need for LTER systems. More broadly, we now have online sites (http://www.kaggle.com/, https://kepler-project.org/) that allow custodians of large-scale compatible data sets to interface with data analysts across the globe, who then compete to provide the best analysis or model from the data. While such approaches once seemed anathema to traditional scientific process, they appear to be necessary for solving complex and multifaceted scientific questions that seemed intractable using conventional studies.

A well-planned LTER system provides a shared scientific infrastructure for all sorts of biodiversity and ecosystem studies, and infrastructure for individual projects within it (Magnusson et al. 2005; Costa and Magnusson 2010; Hero et al. 2010). The RAPELD LTER system allows integration of a diverse range of individual LTER projects using comparable methods. While its design supports LTER projects and publications using conventional scientific research approaches, the system is increasingly demonstrating its capacity to detect and inform managers about black swans.

Acknowledgments The PPBio RAPELD system in Brazil is funded by the Ministry of Science and Technology and many different partners in each Regional Hub (see http://ppbio.inpa.gov.br/Eng/nregionais/). It has received much support from ICMBIO and IBAMA staff, and much of the coordination is undertaken through the National Institute for Science and Technology for Amazonian Biodiversity (INCT-CENBAM). The PPBio system in Australia is partially funded by the Terrestrial Ecosystem Research Network (TERN), Brisbane City Council, SEQ Catchments, Condamine Alliance, Save the Bilby Fund and Griffith University. We thank Sarah Butler, Naomi Edwards, Clay Simpkins, and the many volunteers who measured trees at Karawatha.

References

Abbott I, Le Maitre D (2010) Monitoring the impact of climate change on biodiversity: the challenge of megadiverse Mediterranean climate ecosystems. Austral Ecol 35:406–422

Adis J (ed) (2002) Amazonian arachnida and myriapoda. Pensoft, Sofia

Braga-Neto R, Luizão R, Magnusson WE, Zuquim G, Castilho CV (2008) Leaf litter fungi in a central Amazonian forest: the influence of rainfall, soil and topography on the distribution of fruiting bodies. Biodivers Conserv 17:270–2712

Buckland ST, Magurran AE, Green RE, Fewster RM (2005) Monitoring change in biodiversity through composite indices. Philos Trans R Soc B 360:245–254

Bueno AS, Bruno RS, Pimentel TP, Sanaiotti TM, Magnusson WE (2012) The width of the riparian habitats for understory birds in an Amazonian forest. Ecol Appl 22(2):722–734

Butler S (2007) Associations of mesoscale vegetation patterns with soil, topography and fire history of Karawatha. Hon Thesis, Griffith University, Gold Coast

Calzada J, Delibes M, Keller C, Palomares F, Magnusson WE (2008) First record of the bushy-tailed opossum, *Glironia venusta*, thomas 1912 (Didelphimorphia) from Manaus, Amazonas, Brasil. Acta Amaz 38:807–810

Castilho CV, Magnusson WE, Araújo RNO, Luizão RCC, Luizão FJ, Lima AP, Higuchi N (2006) Variation in aboveground tree live biomass in a central Amazonian forest: effects of soil and topography. For Ecol Manage 42:95–103

Castilho CV, Magnusson WE, Araújo RNO, Luizão F (2010) Short-term temporal changes in tree live biomass in a central Amazonian Forest, Brazil. Biotropica 42:95–103

Cintra R (2008) Aves. In: Oliveira ML, Bacarro FB, Braga-Neto R, Magnusson WE (eds) Reserva ducke: a biodiversidade amazônica através de uma grade. Attema Design Editorial, Manaus, pp 77–85

Condrati LH (2009) Distribuição de espécies de anuros em zonas riparias. Masters Thesis, Instituto Nacional de Pesquisas da Amazônia, Manaus

Costa FRC, Magnusson WE (2010) The need for large-scale, integrated studies of biodiversity—the experience of the program for biodiversity research in Amazônia. Nat Cons 8:1–5

Costa FRC, Magnusson WE, Luizão RC (2005) Mesoscale distribution patterns of Amazonian understory herbs in relation to topography, soil and watersheds. J Ecol 93:863–878

Costa FRC, Castilho CV, Drucker DP, Kinupp V, Nogueira A (2008) Flora. In: Oliveira ML, Bacarro FB, Braga-Neto R, Magnusson WE (eds) Reserva ducke: a biodiversidade amazônica através de uma grade. Attema Design Editorial, Manaus, pp 21–30

Costa FRC, Guillaumet J-L, Lima A, Pereira OS (2009) Gradients within gradients: the mesoscale distribution patterns of palms in a central Amazonian forest. J Veg Sci 20:1–10

Dias MS, Magnusson WE, Zuanon J (2010) Reduced-impact logging has subtle medium-term effects on fish assemblages in central Amazonia. Cons Biol 24:278–286

Doak DF, Estes JA, Halpern BS, Jacob U, Lindberg DR, Lovvorn J, Monson DH, Tinker MT, Williams YM, Wootton JT, Carroll I, Emmerson M, Micheli F, Novak M (2008) Understanding and predicting ecological dynamics: are major surprises inevitable? Ecology 89:952–961

Drucker D, Costa FRC, Magnusson WE (2008) How wide is the riparian zone of small streams in tropical forests? a test with terrestrial herbs. J Trop Ecol 24:65–74

Espírito-Santo HMV, Magnusson WE, Zuanon J, Mendonça F, Landeiro VL (2009) Seasonal variation in fish assemblage composition of small Amazonian forest streams: evidence for predictable changes. Freshw Biol 54:536–548

Eymann J, Degreef J, Häuser C, Monje JC, Samyn Y, VandenSpiegel D (eds) (2010) Manual on field recording techniques and protocols for all taxa biodiversity inventories (ATBIs). Abctaxa 8(1, 2). http://www.abctaxa.be/volumes/volume-8-manual-atbi

Ferraz G, Marinelli CE, Lovejoy TE (2008) Biological monitoring in the Amazon: recent progress and future needs. Biotropica 40:7–10

Field SA, Tyre AJ, Possingham HP (2005) Optimizing allocation of monitoring effort under economic and observational constraints. J Wildl Manag 69:473–482

Gardner T (2010) Monitoring forest biodiversity. Earthscan, London

Goralewski KBN (2008) A influência do delineamento amostral nas estimativas de riqueza e composição de espécies de anfíbios nas margens do alto rio madeira (Rondônia, Brasil). Masters Thesis, Instituto Nacional de Pesquisas da Amazônia, Manaus

Haughland DL, Hero J-M, Schieck J, Castley JG, Boutin S, Sólymos P, Lawson BE, Holloway G, Magnusson WE (2010) Planning forwards: biodiversity research and monitoring systems for better management. Trends Ecol Evol 1210:199–200

Hero J-M, Magnusson WE, Rocha CFD, Catterall CP (2001) Antipredator defenses influence the distribution of amphibian prey species in the central Amazon forest. Biotropica 33:131–141

Hero J-M, Castley JG, Malone M, Lawson B, Magnusson WE (2010) Long-term ecological research in Australia: innovative approaches for future benefits. Aust Zool 35:216–228

Kinupp VF, Magnusson WE (2005) Spatial patterns in the understorey shrub genus *Psychotria* in central Amazonia: effects of distance and topography. J Trop Ecol 21:1–12

Lefsky MA, Cohen WB, Harding DJ, Parker GG, Acker SA, Gower T (2002) LIDAR remote sensing of above-ground biomass in tree biomes. Glob Ecol Biogeogr 11:393–399

Lindenmayer DB, Likens GE (2009) Adaptive monitoring: a new paradigm for long-term research and monitoring. Trends Ecol Evol 5:482–486

Magnusson WE, Lima AP, Hero J-M, Araújo MC (1999) The rise and fall of *Hyla boans*: reproduction in a neotropical gladiator frog. J Herptol 33:647–656

Magnusson WE, Lima AP, Luizão R, Luizão F, Costa FRC, Castilho CV, Kinupp VF (2005) RAPELD: A modification of the gentry method for biodiversity surveys in long-term ecological research sites. Biota Neotrop 2:1–6

Magnusson WE, Costa F, Lima A, Baccaro F, Braga-Neto R, Romero RL, Menin M, Penha J, Hero J-M, Lawson BE (2008) A program for monitoring biological diversity in the Amazon: an alternative perspective to threat-based monitoring. Biotropica 40:409–411

Mendes Pontes AR, Sanaiotti T, Magnusson WE (2008) Mamíferos de grande porte. In: Oliveira ML, Bacarro FB, Braga-Neto R, Magnusson WE (eds) Reserva Ducke: a biodiversidade amazônica através de uma grade. Attema Design Editorial, Manaus, pp 51–56

Mendes Pontes ART, Paula MD, Magnusson WE (2012) Low primate diversity and abundance in northern Amazonia and its implications for conservation. Biotropica 44(6):834–839

Mendonça FP, Magnusson WE, Zuanon J (2005) Relationships between habitat characteristics and fish assemblages in small streams of central Amazonia. Copeia 4:750–763

Mendonça F, Pazin V, Espírito-Santo H, Zuanon J, Magnusson WE (2008) Peixes. In: Oliveira ML, Bacarro FB, Braga-Neto R, Magnusson WE (eds) Reserva ducke: a biodiversidade amazônica através de uma grade. Attema Design Editorial, Manaus, pp 63–75

Menin M, Lima AP, Magnusson WE, Waldez F (2007) Topographic and edaphic effects on the distribution of terrestrially reproducing anurans in central Amazonia: mesoscale spatial patterns. J Trop Ecol 23:539–547

Menin M, Waldez F, Lima AP (2008a) Temporal variation in the abundance and number of species of frogs in 10,000 ha of a forest in central Amazonia, Brazil. South Am J Herpetol 3:68–81

Menin M, Lima AP, Rodrigues D, Waldez F (2008b) Sapos. In: Oliveira ML, Bacarro FB, Braga-Neto R, Magnusson WE (eds) Reserva ducke: a biodiversidade amazônica através de uma grade. Attema Design Editorial, Manaus, pp 21–30

Oliveira PY, Pereira de Souza JL, Baccaro FB, Franklin E (2009) Ant species distribution along a topographic gradient in a "terra-firme" forest reserve in central Amazonia. Pesq Agropec Bras 44:852–860

Parr TW, Ferretti M, Simpson IC, Forsius M, Kovács-Láng E (2002) Towards a long-term integrated monitoring programme in Europe: network design in theory and practice. Env Monit Assess 78:253–290

Pazin VF, Magnusson WE, Zuanon J, Mendonça FP (2006) Fish assemblages in temporary ponds adjacent to 'terra-firme' streams in central Amazônia. Freshw Biol 51:1025–1037

Pitman NCA, Widmer J, Jenkins CN, Stocks G, Seales L, Paniagua F, Bruna EM (2011) Volume and geographical distribution of ecological research in the andes and the Amazon, 1995–2008. Trop Cons Sci 4:64–81

Ribeiro JELS, Hopkins MJG, Vincentini A, Sothers CA, Costa MAS, Brito JM, Souza MAD, Martins LHP, Lohmann LG, Assunção PACL, Pereira EC, Silva CF, Mesquita MR, Procópio LC (1999) Flora da reserva ducke. INPA-DFID, Manaus, AM

Rodrigues DJ, Lima AP, Magnusson WE, Costa FRC (2010) Temporary pond availability and tadpole species composition in central Amazonia. Herpetologica 66:113–119

Schietti J, Drucker D, Keizer E, Carneiro-Filho A, Magnusson W (2007) Avaliação do uso de dados SRTM para estudos ecológicos na Amazônia central. Anais Online do XIII Simpósio Brasileiro de Sensoriamento Remoto (XIII SBSR), INPE, São José dos Campos, SP

Taleb NN (2007) The black swan. Penguin Books, New York

Toledo JJ, Magnusson WE, Castilho CV (2009) Influence of soil, topography and substrates on differences in wood decomposition between one-hectare plots in lowland tropical moist forest in central Amazonia. J Trop Ecol 25:649–656

Toledo JJ, Magnusson WE, Castilho CV, Nascimento HEM (2011) How much variation in tree mortality is predicted by soil and topography in central Amazonia? For Ecol Manag 262:331–338

Toledo JJ, Magnusson WE, Castilho CV, Nascimento HEM (2012) Tree mode of death in central Amazonia: effects of soil and topography on tree mortality associated with storm disturbances. For Ecol Manag 263:253–261
Westoby M (1991) On long-term ecological research in Australia. In: Risser PG (ed) Long-term ecological research. Wiley, New York, pp 191–209
Wintle BA, Runge MC, Bekessy SA (2010) Allocating monitoring effort in the face of unknown unknowns. Ecol Lett 13:1325–1337

Chapter 13
Who's in Conflict with Whom? Human Dimensions of the Conflicts Involving Wildlife

Silvio Marchini

Abstract Some of the most high-profile wildlife conservation cases in the world have been addressed within the emerging field of human–wildlife conflict (HWC). Although HWC is often defined as any situation where wildlife comes into conflict with humans over common resources, the term HWC has been applied almost exclusively to cases involving charismatic mega-fauna, such as large-bodied herbivores and top predators. What these animals have in common is not the magnitude of the damage they cause or their conservation status, but their power to elicit strong mixed opinions among broad sectors of society, which often results in clashes between groups of people who hold differing values toward these animals and their management. As society becomes more diversified and people hold more varied views on human domination over nature, conflicts involving wildlife will grow in intensity and frequency. In this chapter, I discuss the importance of the human dimensions perspective for effectively understanding and resolving HWC; an approach that goes beyond the traditional ecological and economic considerations about reciprocal negative impacts, by addressing also the complexity of the causal relationship between wildlife damage and human thoughts and actions toward wildlife, and the disagreements between people over wildlife values and management objectives.

13.1 Introduction

A variety of wild animals have caused damage and destruction to human property—and sometimes to human life—for as long as humans and animals have shared the same landscapes and resources. In response, throughout civilization, people have

S. Marchini (✉)
Forest Science Department, Luiz de Queiroz College of Agriculture, University of São Paulo, P.O. Box 09, Piracicaba, SP 13418-900, Brazil
e-mail: silvio@escoladaamazonia.org

L. M. Verdade et al. (eds.), *Applied Ecology and Human Dimensions in Biological Conservation*, DOI: 10.1007/978-3-642-54751-5_13,

killed, captured, or otherwise harmed nuisance wildlife. The encroachment of people into wildlife habitat, the competition between wildlife and agricultural producers, and the ability of wild animals to adapt to human-dominated landscapes, all open the way for negative interactions. Ranchers and farmers have always contended themselves and their industries against wildlife, often times to the detriment of the latter. More recently, urban residents are also increasingly experiencing wildlife damage. Examples of such negative encounters range broadly from an annoying opossum in the attic to elephants raiding subsistence crops, and from bird feces corroding building materials to large carnivores such as wolves, bears, pumas, and jaguars breaking into backyards or killing domestic animals.

In the last two decades, a particular subset of these negative interactions between people and wildlife—that involving charismatic mega-fauna—has been increasingly referred to as human-wildlife conflict (HWC). Although HWC has been broadly defined as "when the behavior of wild animal species poses a direct and recurring threat to the livelihood or safety of a community and, in response, persecution of the species ensues" (Zimmermann et al. 2010), not every species that has a negative impact on or is negatively impacted by humans is treated under the label of HWC. Noncharismatic nuisance animals continue to be handled through the traditional animal damage control approach, whereas HWC has been increasingly used to frame some of the most high-profile wildlife conservation cases in the world (e.g., lion and cheetah in Africa, tiger, and elephant in Asia, jaguar in South America, wolf in North America, lynx and raptor in Europe; Hazzah et al. 2009; Dickman 2010; Marker and Dickman 2004; Karanth et al. 2013; Inskip and Zimmermann 2009; Naughton-Treves and Treves 2005; Rabinowitz 2005; Marchini and Macdonald 2012; Bath et al. 2008; Thirgood and Redpath 2005).

The difference between the bats, small rodents, snakes, frogs, or invertebrates that are simply controlled as pests *versus* the carnivores, primates, mega-herbivores and birds addressed in the growing HWC literature is not necessarily the magnitude of the damage they cause or their conservation status, but rather the fact that the animals in the latter group can elicit strong mixed opinions and feelings among broad sectors of society. These animals can be hated and feared—as much as any pest—in one context or by some people, but also highly regarded for their commercial, recreational, ecological, cultural, scientific, spiritual, esthetic, or simply existence value in another context or by other people.

Conflict involving humans and wildlife is rapidly becoming one of the most widespread issues facing wildlife conservationists and managers today, not only because of the greater overlap between human population and wildlife territory, higher costs of the domestic animals and crops lost to wildlife, and aggravated conservation status of the involved wildlife species, but also because society is becoming more diversified, people are holding more varied views about wildlife, and more people now want to participate in decisions regarding wildlife management. While some social groups safeguard the interests of the people affected by HWC, a growing set of social groups advocate for the involved wildlife or habitat. Nonetheless, the study and mitigation of HWC often focuses on the negative ecological and economic consequences—i.e., wildlife injure or kill game or domestic animals, damage crops, and threaten or kill people—without addressing disagreements between people over wildlife values and management objectives.

In this chapter, I discuss the importance of the human dimensions perspective for effectively understanding and resolving HWC. First, I present a brief overview of current ecological and economic approaches to framing and mitigating HWC, revealing their emphasis on wildlife damage. Second, I address the limitation of these approaches in face of the imprecise linkage between damage and human thoughts and actions toward wildlife, and the inherent disagreements between groups of people—for reasons other than ecological and economic ones—regarding management goals. Third, I introduce human dimensions as a complementary, yet neglected field of study and management to deal with the complex and interdisciplinary nature of HWC.

13.2 Ecology and Economics of Human–Wildlife Conflicts

The study and mitigation of HWC has focused on the negative aspects of the interactions between humans and wildlife. More specifically, it has given attention to the patterns and predictors of damage caused by wildlife; the description of the damage to human life and property, with emphasis on livestock depredation and crop-raiding; the monetary costs associated with damage; the implications of the situation for wildlife conservation; and the prevention of damage and mitigation of monetary losses. Below is a brief overview of these ecological and economic considerations about HWC.

13.2.1 Patterns and Predictors of Damage

Livestock depredation and crop-raiding are ecological events and, as such, can be explained through the detection of spatial patterns. Through the application of spatial analysis, factors leading to wildlife damage can be determined and verified. In the last decades, ecologists have benefited from the rapid advances in data collection and computer technologies—notably, Global Positioning System (GPS)-based techniques, camera trapping, geographic information systems (GIS) and spatial analysis software—to vastly improve their ability to collect and analyze data on depredation and crop-raiding events. Results revealed that wildlife damage is patchily distributed in space and time. Some sites are particularly prone to depredation (the so-called "predation hotspots") or crop-raiding, while other areas nearby are unaffected (Wang and Macdonald 2006).

Comparison of affected and unaffected areas may reveal the underlying causes of damage. Extensive research addressing the factors that predispose livestock to depredation showed that opportunity for contact with wildlife and proximity to cover or other wildlife habitat can be valuable quantitative predictors of damage (Soto-Shoender and Giuliano 2011; Treves et al. 2004). Availability of wild prey may also affect damage. Cavalcanti and Gese (2010), for instance, found that "rainfall, and subsequent water levels on the Brazilian Pantanal, was the main driver of seasonal kill rates by jaguars on cattle and caiman. As water levels increased, predation on caiman increased as caiman became more distributed throughout the landscape. Conversely, as water levels fell, caiman became less plentiful, and cattle were moved out into pastures thereby increasing their availability to more jaguars."

Ecology and technology together can point to the damage management solutions most likely to be effective and allow prediction of when and where future damage is likely to occur, so that management can be targeted most effectively. In order to predict retaliatory persecution, however, managers will have to look beyond the ecology of damage to examine how individuals and communities respond to different patterns of damage. In a landscape dominated by small properties, for example, a patchy distribution of damage may give disproportionate weight to rumors and perceptions of risk, especially if a few properties may be severely and repeatedly affected.

13.2.2 Types and Magnitude of Damage

Predation upon livestock—cattle, horses, sheep, pigs, goats, or poultry—has been the most common type of damage addressed in the study and management of HWC. The problem is widespread, including lion, leopard, striped hyena, and African wild dog in Africa; Eurasian lynx, wolverine, brown bear, and gray wolf in Europe; coyote and gray wolf in the USA (Thirgood et al. 2005); tiger in India (Mishra 1997) and jaguar and puma in Brazil (Rabinowitz 2005; Mazzolli et al. 2002). Depredation can be intense. For instance, Mazzolli et al. (2002) found that losses to pumas in southern Brazil were 84 % for sheep, 78 % for goats, and 16 % for cattle, and villagers in Nepal reported that 63 % of all stock deaths were due to predators (Jackson et al. 1996). Surplus killing, where predators kill multiple animals in one attack, can result in severe financial hardship to the stockowners concerned (Nowell and Jackson 1996) and engenders particularly intense hostility toward carnivores (Jackson 2000). In many cases, however, depredation accounts for a relatively small level of stock offtake, particularly compared with other causes of stock loss. For instance, African wild dogs were found to cause only 1.8 % of stock losses on cattle ranches in Zimbabwe, while disease caused 23.5 % (Rasmussen 1999).

Crop-raiding, another common type of damage associated with HWC, can be simply defined as wild animals moving from their natural habitat onto agricultural land to feed on the products that humans grow for their own consumption. Examples include birds and monkeys alone destroying up to 77 % of a potential crop in Latin America (Perez and Pacheco 2006) and white-tailed deer inflicting millions worth of crop damage annually in the USA (Naughton-Treves and Treves 2005). Elephants are probably the animals most commonly associated with crop-raiding in the HWC literature. They not only trample crops but occasionally kill or injure people too. Nonetheless, studies suggest that small animals such as primates and rodents cause more damage than larger animals in the long run (Naughton-Treves and Treves 2005).

Wildlife damage traditionally has been thought of as just a rural or agriculture problem (Messmer 2000). More recently, though, overabundant wildlife populations have been causing a myriad of other problems, including residential damage and vehicle collisions. Over 60 % of urban and suburban households in the USA annually experience problems with wildlife (Conover 1997). Urban residents also reported spending over 260 million hours trying to solve or prevent these problems (Conover 1997).

Human fatalities and illnesses resulting from interactions with wildlife are less common than damage to property, but far more emotive. They result from wildlife-related diseases, wildlife bites, attacks, automobile collisions, and bird–aircraft strikes. Research suggests that in the USA each year approximately 5,000 people are injured or become ill, and 415 people die because of wildlife-related incidents (Conover 2002). Despite relative rarity worldwide, wildlife attacks on humans can pose a significant threat in some areas: for instance, the Sundarbans region in India has long been a "hotspot" for man-eating tigers, with around 100 human deaths reported annually (Sanyal 1987). Asian elephants kill 100–200 people every year in India (Veeramani et al. 1996). Risks of disease transmission can also lead to hostility toward wildlife. For instance, cases of Brazilian spotted fever have been associated by the public opinion with capybaras and their ticks, and this has resulted in capybaras being blamed for the recent increase in the occurrence of the disease in southeastern Brazil, despite the fact that Brazilian spotted fever can be transmitted by a variety of species, including, by not restricted to, the capybara (Moreira et al. 2012). Mosquitos, snakes, and even domestic dogs, however, cause far more human fatalities than the large carnivores and mega-herbivores of the HWC literature (World Health Organization 2010).

13.2.3 Economic Costs of Damage

Sharing the space with wildlife can incur substantial economic costs. In the USA, for example, urban households lost US$63 per household, or US$1.9 billion total, because of wildlife damage and spent US$5.5 billion to manage wildlife problems during the 1990s, while agricultural producers spent US$2.5 billion over the same period (Bruggers et al. 2002; Conover 1997). In Nepal, depredation by wolves and snow leopards costs villagers around 50 % of their average annual per capita income (Mishra 1997). The costs of maintaining large carnivores can extend much further than the individual farmers. In Norway, for instance, the government paid out more than US$3 million in compensation for stock losses to carnivores in

2000 alone (Swenson and Andren 2005). Wildlife damage can result in a variety of additional costs aside from the direct impact of depredation or crop-raiding, as people have to invest more heavily in strategies such as livestock herding, guarding, and predator or mega-herbivore control (Thirgood et al. 2005). There may also be additional "opportunity costs" associated with the presence of wild animals, as the time required for livestock protection limits the amount of time that can be invested in other potentially important activities such as attending school or assisting with crop harvesting (Barua et al. 2013).

13.2.4 Retaliation to Damage: Conservation Implications

Lethal control of wildlife associated with damage has resulted in dramatic population declines, striking contractions in geographic range, and often local extirpation (Johnson et al. 2001; Treves and Naughton-Treves 2005). Well-documented examples include the African lion, which has suffered a substantial population decline and range contraction over recent decades, and has disappeared from much of its historic range (Nowell and Jackson 1996). The cheetah has also declined from an estimated population of around 100,000 individuals in 1,900 to less than 15,000 today, restricted almost exclusively to small, fragmented populations (Marker 2002). Similarly, after centuries of persecution, African wild dogs remain in only 14 of the 39 countries they once occupied and are now one of the world's most endangered carnivores, numbering fewer than 5,000 individuals worldwide (Woodroffe et al. 1997). Although these declines are often due to multiple factors, including habitat loss, degradation and fragmentation, and disease risks, killing by humans is an increasingly important factor driving declines for many species and is therefore a highly important and increasingly urgent conservation issue (Marker and Dickman 2004; Nowell and Jackson 1996; Zeller 2007).

13.2.5 Resolving the Problem by Changing the Ecological Context

Given that wildlife damage is an ecological phenomenon governed by the opportunity for contact between people (and their property) and wildlife, a reasonable conflict-reduction strategy is to change the ecological context in order to decrease the severity and frequency of encounters between wildlife and people/property. Ways to accomplish this include lethal control, such as regulated hunting in developed countries where legislation is strong, or selective removal of problem animals from human settlements through government intervention, while the less ecologically correct "shooting-shoveling-shutting up," also known as the 3-S treatment, remains as the principal method to resolve wildlife damage problems worldwide. Treves et al. (2009) reviewed the interventions for mitigating HWC by reducing the severity and frequency of encounters between wildlife and people/property and listed them as follows: barriers, guards, repellents, manipulate problematic animals (lethal/permanent), manipulate problematic animals (nonlethal/temporary), manipulate habitat or other wildlife, protect wildlife or habitats, and reduce attractiveness of property/people.

13.2.6 Resolving the Problem by Changing the Economic Context

Compensation payments for livestock lost to predators or crops raided by ungulates are a widespread mitigation strategy used to reduce the economic costs that result from wildlife damage. They are generally viewed as efforts to increase people's tolerance of problem species (Treves et al. 2009). However, compensation schemes usually have unforeseen effects. They may result in a neglect of preventive measures (Nyhus et al. 2005) or make people dependent on payment (Bulte and Rondeau 2005). International donor agencies and NGOs, in response to accusations of neo-imperialism (Brockington 2002), have turned to community-based efforts for conflict mitigation. Generating income and redistributing revenues earned through ecotourism have been promoted as an alternative. However, its efficacy has been questioned (Kiss 2004). For example, Hemson et al. (2009) point out that benefits from ecotourism are unevenly shared, while the costs of human–wildlife conflict are widespread.

13.3 When Ecology and Economics are not Enough

Ecology and economics provide a wide array of tools and techniques for understanding and managing wildlife damage and conservation issues. Together, these disciplines have contributed significantly to the control of pests and the conservation of endangered species. However, species involved in HWC are not necessarily endangered and definitely not seen as pests. They are charismatic mega-fauna, and charisma, alongside other subjective values, does not belong in the realms of ecology and economics. In the interactions between humans and charismatic animals, the cause–effect relationship between wildlife damage and negativity toward wildlife is seldom simple and consistent. Besides, people often disagree—based on values other the ecological and economic ones—about management goals in HWC: while some people favor the control of damage to the detriment of wildlife, others favor wildlife for its positive impacts. In this section, I discuss two facts that can render ecology and economics insufficient to resolve HWC: wildlife damage alone does not necessarily explain human behavior toward wildlife (e.g., persecution), and people often disagree about goals and alternatives regarding wildlife management.

13.3.1 Damage Alone does not Explain Human Actions Toward Wildlife

In HWC, persecution (i.e., persistent killing, chasing, or other harassment of a species) is not always a simple function of wildlife damage. The reason for this is twofold. First, there may be a discrepancy between actual and perceived damage. What ultimately drives human behavior is not the reality itself, but how reality is perceived. In conflicts with high-profile animals such as top predators and mega-herbivores, the perceived damage and risk often exceed the actual evidence (Conover 2002; Marchini and Macdonald 2012). Feedback loop between notoriety and sensationalism may be behind the distorted perceptions. For instance, Marchini (2010) found that jaguar attacks on people have a prominent place in story-telling in rural Brazil and the attack reports provided by respondents may have been inflated by cognitive biases such as availability cascade [i.e., a self-reinforcing process in which a collective belief gains more plausibility through its increasing repetition in public discourse (colloquially, "repeat something often enough and it will become fact")] (Kuran and Sunstein 1999) or availability heuristic (i.e., answers depend on what is more available in memory, which is biased toward vivid, unusual, or emotionally charged episodes such as a jaguar attack) (Tversky and Kahneman 1974). Furthermore, antagonism toward wildlife can persist for many years; in part of the northern Ethiopian highlands, people were negative toward leopards said leopards killed livestock, even though leopards no longer occurred locally (Yirga et al. 2011), and Macdonald (1987) describes similarly how accounts behind the conflicts serious attacks by foxes on lambs in the north of England often turned out to refer to the folkloric experiences of other people long ago.

Second, factors not directly related to the impacts that wildlife have on human livelihoods may also be involved in the persecution of wildlife. Large carnivores, for example, elicit strong negative emotions, particularly fear (Manfredo 2008), with people who are fearful of carnivores usually being more antagonistic to them (Roskaft et al. 2007). Fear is likely to have evolved as a heritable anti-predator response and can be exaggerated in the light of fact; for instance, a 2002 study in Norway revealed that 48 % of respondents were very afraid of wolves, with an additional 40 % somewhat afraid, despite the fact that only one documented wolf attack has ever occurred in Norway, in 1800 (Linnell et al. 2003). Persecution to wildlife can result also from social motivations. Marchini (2010) found that social motivations are important determinants of the intention to kill jaguars in the Pantanal, where 25 % of ranchers justified their approval of jaguar killing on the grounds of tradition. These ranchers often refer, with apparent pride, to the "*Pantaneiro* culture" and the conviction that jaguar hunting has been passed from generation to generation as an element of that culture. Likewise, Hazzah et al. (2009) found that social identity was behind the killing of carnivores by traditional pastoralist groups such as the Maasai; killing lions is central to their culture, young warriors are expected to kill lions and are celebrated when they do so. Dickman et al. (2013) discuss how other factors at the individual level (e.g., experience, skills, knowledge, and values) and societal/cultural level (e.g., income sources, folklore, and religion) affect human behavior toward wildlife (e.g., wildlife killing).

13.3.2 Social Groups Disagree Over Goals and Management Alternatives

Animals involved in HWC evoke strong, mixed opinions, and feelings. Nobody opposes the extermination of mosquitos or gets offended by the nonconsumptive use of birds in bird-watching, but the management of iconic animals such as lions, wolves, and elephants divides opinions among broad sectors of society, which can result in social conflict. Damage alone can explain disagreements between groups of people when the impact of HWC is differentially distributed, with people more heavily affected expecting more stringent measures against wildlife than those who are not negatively affected. Social conflicts over wildlife, however, are often based on subjective factors. People can disagree over goals and management alternatives for affective, esthetic, or ethical reasons, for example. While the loss of livestock and crops to carnivores and large-bodied herbivores is something tangible, immediate, objectively measurable, and therefore amenable to rational analyses and negotiation, emotional, esthetic, and ethical values are subjective, varying across individuals, social segments, and cultures, and rendering ineffective the mitigation measures that are based on the logic of ecology and economics. Ironically, within the most influential social segment involved in conflicts over wildlife—that of the conservation professionals—the prevalent wildlife value is probably one of the most subjective and hard to communicate; existence value, which is defined as the benefit people receive from simply knowing that the wildlife in question exists.

Moreover, wildlife often becomes surrogate for deeply embedded cultural discords within and between social groups. The dispute over wolves in Yellowstone National Park, for example, was found to be indicative of a broader ideological debate over property and natural resources management (Wilson 1997). Likewise, social disagreements over the management of exotic or "foreign" birds in Western Europe were found to be connected to the public debate over the influx of human immigrants from Eastern Europe (Fine and Christoforides 1991). Conflicts between social groups over wildlife can be aggravated by the urban–rural divide. Urbanization generates wildlife advocacy, but the immediate costs of living with wildlife are (or are perceived to be) borne by the rural populations (Swenson and Andrén 2005). There is an exceptionally high rate of urbanization (1.8 %) among the world's already highly urbanized countries, such as Brazil (86 % of the total population living in urban areas) (Central Intelligence Agency 2010). Ranchers and farmers are a minority group in Brazil, and their numbers are dwindling. As a result, they may associate wildlife conservation with urban values that are increasingly imposed on them and might view the continuation of wildlife killing as part of their resistance to this and their struggle to preserve their rural heritage (Marchini 2010). Protection of this heritage underlies also conflicts over conservation of prairie dogs in North America's prairies (Reading et al. 2005) and large carnivores in Norway (Swenson and Andrén 2005).

Hence, negativity toward wildlife in HWC is not merely determined by any direct costs imposed, but is rather the product of a dynamic and complex web of individual (e.g., perception of risk and fear), societal (e.g., peer pressure and social conflict), and cultural (e.g., identity) factors. The imprecise linkage between actual wildlife damage and wildlife killing may turn irrelevant many biologically based conservation actions and mitigation measures, which can end up adding a potentially lethal element to already significant risks to a threatened species posed by retributive kill, or aggravating already existing tensions between affected social segments and wildlife professionals.

Nonetheless, HWC has been often addressed from the traditional wildlife management perspective, for which "the most important task is to choose the right goal and to know enough about the animals and their habitat to assure its attainment" (Sinclair et al. 2006). Although wildlife management has been historically successful at attaining its four goals regarding a wildlife population, namely "make it increase, make it decrease, harvest it for a continuing yield, and leave it along but keep an eye on it" (Sinclair et al. 2006), the proposed focus on animals and their habitat excludes an explicitly obligatory element of human–wildlife conflicts: humans. Sinclair et al. (2006) themselves admit that of the three decisions that are needed to attain management goals—(1) what is the desired goal; (2) which management option is therefore appropriate; and (3) by what action is the management option best achieved—the two latter require technical judgment but the first decision requires a judgment of value. The very essence of HWC is the disagreement about value, and this renders the task of choosing "the right goal" in conflict situations particularly challenging for the traditional wildlife professional.

13.4 Human Dimensions of the Conflicts Involving Wildlife

The biological tradition of the wildlife profession, with its emphasis on animals and their habitats, has proved inadequate to deal with the social nature of many of the current conservation and management problems, notably human–wildlife conflicts. In the USA, for example, where human–wildlife interactions and the stakeholders of wildlife management increased and diversified significantly over the last three decades, the wildlife profession gradually expanded its scope to incorporate a "human dimensions" perspective. Human dimensions emerged in that country as a sub-discipline of wildlife management and today it is understood as "a field of study that applies social sciences to examine human–wildlife relationships, and in doing so, provides information that contributes to effective wildlife conservation efforts" (Manfredo 2008).

Wildlife management from the human dimensions perspective addresses the system human-wildlife-habitat, instead of wildlife and habitat separately, with an emphasis on describing, understanding, predicting, and affecting human though and action toward wildlife (Manfredo et al. 1996). As humans are the common thread in the highly variable realm of human–wildlife conflicts, and the course and resolution of conflict are determined by the thoughts and actions of the people involved, understanding the human dimensions is the most crucial prerequisite for developing effective mitigation (Manfredo and Dayer 2004). Three concepts in human dimensions are particularly useful for the study and mitigation of HWC; at the individual level, "impacts" determine a person's tolerance and behaviors toward wildlife; at the social group level, the identification and engagement of "stakeholders" are necessary steps toward the mitigation of social conflicts over wildlife; and by integrating impacts and stakeholders into "capacity" concepts, managers can determine the size of the wildlife population that produces the best overall impact to society.

13.4.1 Impacts

Impacts are thought as the subset of the effects generated by the interactions between humans, wildlife, and wildlife habitat that are recognized by the people involved, and interpreted as being important; important enough to draw management attention (Riley et al. 2002). If an effect of human wildlife interaction does not require management attention, it is not an impact. Impacts take a variety of forms, so it is useful to organize impacts into a manageable number of categories. Impacts can be thought, for instance, as economic and ecological, but also cultural, social, psychological, and related to health and safety. Unlike the concept of impact in the strict contexts of ecological and economic theories, impacts in human dimensions of wildlife management are not objectively assessed, but rather defined and weighted by human values.

People evaluate impacts as negative or positive, "bad" or "good." Examples of negative impacts are the frustration of loosing livestock to predators, the fear of getting a disease from the bats in the attic, and the hatred toward the pigeons in the park. Much of wildlife damage management involves minimizing the negative ("bad") impacts associated with wildlife. But positive impacts also play a relevant role in human–wildlife relationships. Examples of positive values are the increased yield that results from the pollination service provided by bees and the enjoyment associated with watching birds in the backyard. A particular interaction between wildlife and people may cause both negative and positive impacts, and different stakeholders can have different evaluations of the same interaction. Even the same person may perceive an interaction as causing both negative and positive impacts simultaneously. Whether that individual evaluates the overall interaction as negative or positive depends on how he or she weighs the importance of each negative and positive impact.

Understanding and influencing the way people define and weigh impacts are at the core of human dimensions of wildlife management programs. A variety of social sciences provide useful information for this, but the discipline relied upon most frequently is social psychology (Decker et al. 2012). Social psychology offers wildlife managers insight into the basis for people's perceptions of impacts because impacts typically are based on values but expressed in terms of attitudes (i.e., favorable or unfavorable dispositions toward a person, an object, an action, etc.; attitude is not synonym to action) and preferences. Examining the concepts underlying the process of human thought to action, such as values, beliefs, risk perception and acceptance, attitudes and norms, and understanding the relationships among them, can be a basis for common management interventions such as communication, education, and incentives.

Theories in social psychology, such as the theory of planned behavior (Ajzen 1985), propose that human behaviors are governed not only by personal attitudes, but also by social pressures and perceived control over one's own behavior. These theories help explain variations in people's support to management actions. For example, lethal control of jaguars was explained by attitudes toward jaguar killing and social pressure among ranchers in the Brazilian Pantanal, whereas among immigrants on the Amazon deforestation, it was explained by attitudes and perceived ease or difficulty of persecuting jaguars (Cavalcanti et al. 2010). Carter et al. (2012) emphasize the importance of psychological frameworks for wildlife conservation and propose a model that integrates the expansive and generalized set of psychological concepts.

Other useful social science disciplines in human dimensions are sociology and economy. Sociology is concerned with how a person's values, norms, attitudes, and other cognitions are influenced by the society or social structure in which he or she lives and interacts. It addresses the factors that account for differences between people in diverse social and cultural conditions and can help managers identify similarities and differences among situation involving people and wildlife in different countries or regions and suggest whether techniques found successful in one area are likely to work in another. Economics, on the other hand, can help managers understand the flow of wildlife values—usually measured in monetary terms—through society, and express its impacts in terms of costs and benefits. Even for attributes of wildlife that cannot be assigned a market value, economists have developed ways of nonmonetary and nonconsumptive valuation that are applicable to wildlife management (Decker et al. 2012).

Negative impact	Positive impact: Non-significant	Positive impact: Significant
Significant	a. Mosquitos, pigeons, rats b. Make it decrease c. Lethal control	a. Bears, jaguars, elephants b. Make it decrease vs. make it increase/harvest it for a continuing yield c. Lethal control vs. protection
Non-significant	a. Majority of species b. No management action needed	a. Honey bees, deer, whales b. Make it increase/harvest it for a continuing yield c. Consumptive and non-consumptive use

Fig. 13.1 Wildlife management based on impacts **a** example of wildlife, **b** management goal (Sinclair et al. 2006), **c** example of management action

An impacts approach to HWC provides managers with a more balanced view of the conflict situation, one that goes beyond the conventional focus on damage. Actually, in light of the impacts approach, HWC can be thought as an interaction that generates both significant negative and positive impacts at the same time (Fig. 13.1). While management goals can be more easily set when impacts are either predominantly negative or predominantly positive ("make it decrease" and "make it increase or harvest it for a continuing yield," respectively), the impacts-based approach can help managers deal with the complexity of HWC situations by identifying the antagonistic values at the core of the conflict.

To illustrate how the assessment of impacts can help managers understand HWC situations and design more effective strategies to deal with them, consider the following hypothetical situations: (1) a predator kills a domestic animal, but the animal's owner (hereafter called rancher) does not know, (2) the predator kills the domestic animal, the rancher knows that, but does not care, (3) the predator kills the domestic animal, the rancher knows and cares about that, but does not do anything, (4) the predator kills the domestic animal, the rancher cares about that and retaliates killing the predator, and (5) the predator does not kill any domestic animal, but the rancher kills the predator.

Within the traditional approach to HWC, with focus on damage and persecution, all five situations can be considered as HWC. In fact, the term HWC has been used to refer to both carnivore predation on livestock and carnivore killing by humans as if depredation and persecution were synonyms to HWC. An analysis of impacts, however, reveals the fundamental differences between these situations and points to the proper management goals. In situations (1) and (2), there is no impact, and therefore no conflict, since the rancher does not have a negative evaluation of the interaction, therefore, management is not needed (though a wildlife ecologist knowing of the depredation event from his or her telemetry study, and without asking first the livestock's owner what he thinks, would call it HWC). The situation (3) is not of conservation concern, given that the rancher does not kill the predator, but the rancher's anger toward the predator calls for management action: "make it decrease" (at least in the ranch). Situation (4) is the classic two-way negative interaction, of both conservation and management concerns. From the rancher's perspective, the management goal is "make it decrease." In the real world, however, social segments that perceive a positive impact from their interaction with the predator (e.g., the enjoyment of knowing that the predator exists) could oppose the population control, which would result in social conflict over management goals. Finally, in the situation (5), there can be a positive impact if the rancher is a sport hunter and kills the predator for pleasure. There is no conflict between the rancher and predator. The management goal would be "make it increase" or "harvest it for a continuing yield," but in the real world, social groups that advocate for wildlife could oppose the management goals supported by the rancher, generating a social conflict.

13.4.2 Stakeholders

Stakeholder is defined as any person or group who significantly affects or is significantly affected by wildlife management (Decker et al. 1996). Identifying the stakeholders is a crucial first step to understanding a social conflict over wildlife. According to Decker et al. (2002), the basic rule in identifying stakeholders is that anyone who is affected by or who can affect management is a valid stakeholder; people concerned about wildlife-related problems and people concerned about the welfare of wildlife; supporters and opponents of management agencies. All people have a right to have their voice heard in decisions that affect them. Therefore, it is imperative that managers do not exclude stakeholders with whom they disagree. The stakeholders who are most interested in an issue often are easiest to identify. Some initiate contact with managers, requesting information or offering opinions. Stakeholder groups can organize to promote their common interests in a particular issue. Examples of these groups include associations of agricultural producers, wildlife users' (e.g., wildlife-based tourist guides), and animal welfare organizations.

Once the stakeholders are identified, they have to be engaged in the conflict management decisions. Engagement means involving people in making, understanding, implementing, or evaluating wildlife management decisions (Decker et al. 1996). Though strategies for effective stakeholder engagement vary by context, it is helpful to consider some general steps in the design of an engagement process (Decker et al. 2002). First, it helps to develop a situation analysis to describe the characteristics of the HWC situation to be managed. Clear objectives for involving stakeholders in conflict management need to be established. Objectives might include improving the information base for decision making, improving the judgments on which decisions are based, or improving the social environment in which conflict management occurs. Once these steps are taken, the manager will be better able to select an overarching stakeholder engagement approach. Selecting the best approach will depend upon a variety of factors, including the level of conflict over the issue, the number and type of stakeholders affected, stakeholder interest in and awareness of the issue, the existence of other entities that can influence management, resource limitations, and the need for information from stakeholders.

13.4.3 Carrying Capacity

A human dimensions approach that takes impacts and stakeholders into account can help managers integrate biological limits with social, economic, institutional, administrative, cultural, and legal limits. Stakeholders vary widely in what positive and negative impacts they experience from wildlife and in their perceptions of the size of the wildlife populations that produces desirable levels of such impacts. While carrying capacity in wildlife management is most often viewed in the classical ecological sense of biological carrying capacity (i.e., the natural limit of a wildlife population), a human dimensions perspective on carrying capacity is based on the assumption that bounds exist on the impact that stakeholders will accept. It acknowledges that they will tolerate negative impacts associated with wildlife only to a certain point beyond which wildlife become intolerable. Concepts such as cultural carrying capacity (Ellingwood and Spignesi 1986), wildlife acceptance capacity (Decker and Purdy 1988), social carrying capacity (Minnis and Peyton 1995), and wildlife stakeholder acceptance capacity (Carpernter et al. 2000) were developed to consider differences in stakeholders' tolerance for impact levels.

13.4.4 Integrating Biological and Social Sciences into Decision Making

Given the complex relationships between wildlife damage and human actions toward wildlife, and the diverse range of stakeholders who desire involvement in decision making, conflict management is no longer primarily a technical problem of mitigating damage. Conflict management issues are, indeed, more than just complex, they are also "wicked" (Decker et al. 2012). "Wicked problem" is a phrase used to describe a problem that is difficult or impossible to solve because of scientific uncertainty about cause–effect relationships and social conflicts over goals and management alternatives. The term "wicked" is used, not in the sense of evil but rather its resistance to resolution. Moreover, because of complex interdependencies, the effort to solve one aspect of a wicked problem may reveal or create other problems (Decker et al. 2012). Human–wildlife conflicts are wicked problem par excellence.

Decision making in HWC, therefore, cannot happen as discrete events or in a linear process, but rather in a cyclic, iterative, dynamic, and adaptive process. Structured Decision Making (SDM) offers a framework for this (Martin et al. 2009). SDM combines analytical methods drawn from the decision sciences and applied ecology with deliberative insights into cognitive psychology, facilitation and negotiation, in a way that is rigorous, inclusive, defensible, and transparent. The value of this process for addressing major challenges in conservation conflict management is that it helps in setting realistic goals; entails a transparent decision-making process, and addresses differing world views and patchy or contested information (Redpath et al. 2013).

13.5 Conclusion

The term human–wildlife conflict has been used to refer to two different situations: one where wildlife injure or kill game or domestic animals, damage crops, and threaten or kill people (with or without preventive or subsequent retaliatory wildlife killing), and one where groups of people disagree about goals and alternatives regarding wildlife management. The former situation is ultimately an ecological phenomenon. Indeed, the dispute between humans and wildlife for resources (e.g., livestock, crops, game, space), the attacks of wildlife on humans, and the transmission of diseases from wildlife to humans or livestock can be understood within the well-established theoretical frameworks of community ecology (e.g., competition, predator–prey, and host–pathogen). "Conflict," on the other hand, does not belong to the glossary of ecological terms. The latter situation is the one of actual conflict, defined as "a situation of competition in which the parties are aware of the incompatibility of potential future positions and in which each party wishes to occupy a position that is incompatible with the wishes of the other" (Boulding 1963). Because wildlife is unlikely to be aware and wishful of a future condition, conflict—as defined above—is exclusively human. The two situations are closely related and may happen at the same time, but clumping them together under the term "human–wildlife conflict" may constrain the way problems are defined and limit the array of potential solutions available. Methods used to resolve wildlife damage problems, for example, differ from the solutions to social conflicts.

The human dimensions approach is vital to resolving the social conflicts behind the problems that have been referred to as human–wildlife conflict. Nonetheless, human dimensions have been poorly researched and insufficiently represented in the action plans and public policies for wildlife management and conservation. Two factors contribute to this discrepancy. First, the trend in conservation toward modes of management that emphasize the landscape scale. Benefiting from advances in data collection and computer technology, spatial modeling and management at the landscape level, yet necessary and of great value for systematic conservation planning, may ignore intimate scales of analysis that are needed to address people's thoughts and actions. Second, human dimensions consist in an interdisciplinary approach that combines perspectives from the ecological and social sciences, and interdisciplinary training within wildlife management and conservation is limited.

The integration of human dimensions into wildlife and natural resources management can be particularly beneficial in developing countries with high biodiversity such as Brazil, where the combination of economic growth, mounting pressure on natural resources, urbanization, expanding human settlement, and agricultural frontier in some regions and wildlife repopulating human-dominated landscapes in other regions, along with the growing ideals of democracy accompanied by greater participation in governance by a growing set of stakeholders, is likely to generate intense conflicts over endangered species as well as natural resources (e.g., water). Capacity building in human dimensions of wildlife and natural resources should be a priority in these countries, so that interdisciplinary, more effective approaches to HWC and biological conservation that integrate ecological and social sciences, can be properly incorporated into research, conservation, management, and public policy.

All the ecological and sociological science in the world, however, does not convey to a wildlife manager what should be done in a given situation. Science informs managers of what is desired by stakeholders, what can be done, and what may happen with and without a particular intervention. The question of what should be done, however, requires ethical considerations. Human dimensions insights help wildlife professionals consider ethical dilemmas by clarifying pertinent values in a wildlife management issue (Decker et al. 2012). Also, by clarifying how society values wildlife, human dimensions can help manager to shift the focus from aiming to maximize wildlife populations to the more difficult, yet more promising one of aiming to optimize wildlife values for society (Messmer 2009).

References

Ajzen I (1985) From intentions to actions: a theory of planned behavior. In: Kuhl J, Beckman J (eds) Action-control: from cognition to behavior. Springer, Heidelberg, pp 11–39

Barua M, Bhagwat SA, Jadhav S (2013) The hidden dimensions of human–wildlife conflict: health impacts, opportunity and transaction costs. Biol Conserv 157:309–316

Bath A, Olszanska A, Okarma H (2008) From a human dimensions perspective, the unknown large carnivore: public attitudes toward Eurasian Lynx in Poland. Hum Dimensions Wildl 13:31–46

Boulding K (1963) Conflict and defense: a general theory. Harper & Brothers, New York
Brockington D (2002) Community conservation, inequality and injustice: myths of power in protected area management. Conserv Soc 2:1–22
Bruggers RL, Owens R, Hoffan T (2002) Wildlife damage research needs: perceptions of scientists, wildlife managers, and stakeholders of the USDA/Wildlife Services program. Int Biodeterior Biodegrad 49:213–223
Bulte EH, Rondeau D (2005) Why compensating wildlife damages may be bad for conservation. J Wildl Manag 69:14–19
Carpenter LH, Decker DJ, Lipscomb JF (2000) Stakeholder Acceptance Capacity in Wildlife management. Hum Dimens Wildl 5:5–19
Carter NH, Riley SJ, Liu J (2012) Utility of a psychological framework for carnivore conservation. Oryx 46:525–535
Cavalcanti SMC, Gese EM (2010) Kill rates and predation patterns of jaguars (*Panthera onca*) in the southern Pantanal, Brazil. J Mammal 91:722–736
Cavalcanti SMC, Marchini S, Zimmermann A, Gese EM, Macdonald DW (2010) Jaguars, livestock and people in Brazil: reality and perceptions behind the conflict. In: Macdonald DW, Loveridge AJ (eds) Biology and conservation of wild felids. Oxford University Press, Oxford, pp 383–402
CIA (2010) The world factbook. Central Intelligence Agency. https://www.cia.gov/library/publications/the-world-factbook. Cited 20 Apr 2010
Conover M (2002) Resolving human–wildlife conflicts: the science of wildlife damage management. CRC Press, Boca Raton
Conover MR (1997) Wildlife management practices by metropolitan residents in the United States: practices, perceptions, costs, and values. Wildl Soc Bull 25:306–311
Decker DJ, Purdy KG (1988) Toward a concept of wildlife acceptance capacity in wildlife management. Wildl Soc Bull 16:53–57
Decker DJ, Lauber TB, Siemer WF (2002) Human–wildlife conflict management: a practitioner's guide. Human Dimensions Research Unit, Cornell University, Ithaca
Decker DJ, Riley SJ, Siemer WF (2012) Human dimensions of wildlife management, 2nd edn. Johns Hopkins University Press, Baltimore
Decker DJ, Krueger CC, Baer Jr RA, Jr., Knuth BA, Richmond ME (1996) From clients to stakeholders: a philosophical shift for fish and wildlife management. Hum Dimens Wildl 1:70–82
Dickman AJ (2010) Complexities of conflict: the importance of considering social factors for effectively resolving human–wildlife conflict. Anim Conserv 13:458–466
Dickman AJ, Marchini S, Manfredo M (2013) The importance of the human dimension in addressing conflict with large carnivores. In: Macdonald DW, Willis KJ (eds) Key topics in conservation biology, vol 2. Wiley-Blackwell, Oxford, pp 110–126
Ellingwood MR, Spignesi JV (1986) Management of an urban deer herd and the concept of cultural carrying capacity. Trans Northeast Deer Tech Comm Vt Fish Dep 22:42–45
Fine GA, Christoforides L (1991) Dirty birds, filthy immigrants and the English sparrow war: metaphorical linkage in constructing social problems. Symb Interact 14(4):68–88
Hazzah L, Borgerhoff-Mulder M, Frank LG (2009) Lions and warriors: social factors underlying declining African lion populations and the effect of incentive-based management in Kenya. Biol Conserv 142:2428–2437
Hemson G, Maclennan S, Mills G, Johnson P, Macdonald DW (2009) Community, lions, livestock and money: a spatial and social analysis of attitudes to wildlife and the conservation value of tourism in a human–carnivore conflict in Botswana. Biol Conserv 142:2718–2725
Inskip C, Zimmermann A (2009) Human–felid conflict: a review of patterns and priorities worldwide. Oryx 43:18–34
Jackson RM (2000) Snow leopard. In: Reading RP, Miller B (eds) Endangered animals: a reference guide to conflicting issues. Greenwood Press, Westport, pp 259–266
Jackson RM, Ahlborn C, Gurung M, Ale S (1996) Reducing livestock depredation losses in the Nepalese Himalaya. In: Proceedings of the 17th Vertebrate Pest Conference. University of California, Davis, pp 241–247

Johnson WE, Eizirik E, Lento GM (2001) The control, exploitation and conservation of carnivores. In: Gittleman JL, Funk SM, Macdonald DW, Wayne RK (eds) Carnivore conservation. Cambridge University Press, Cambridge, pp 196–219
Karanth K, Gopalaswamy A, Prasad P, Dasgupta S (2013) Patterns of human–wildlife conflicts and compensation: insights from Western Ghats protected areas. Biol Conserv 166:175–185
Kiss A (2004) Is community-based ecotourism a good use of biodiversity conservation funds? Trends Ecol Evol 19:232–237
Kuran T, Sunstein CR (1999) Availability cascades and risk regulation. Stanford Law Rev 51(4):123–134
Linnell JDC, Solberg EJ, Brainerd S (2003) Is the fear of wolves justified? A Fennoscandian perspective. Acta Zool Lituanica 13:34–40
Macdonald DW (1987) Running with the Fox. Unwin Hyman, London
Manfredo MJ (2008) Who cares about wildlife? Springer, Berlin
Manfredo MJ, Dayer A (2004) Concepts for exploring the social aspects of human–wildlife conflict in a global context. Hum Dimens Wildl 9(4):1–20
Manfredo MJ, Vaske JJ, Sikorowski L (1996) Human dimensions of wildlife management. In: Ewert AW (ed) Natural resource management: the human dimension. Westview Press, Boulder, pp 53–72
Marchini S (2010) Human dimensions of the conflicts between people and jaguars (*Panthera onca*) in Brazil. Department of Zoology, University of Oxford, Oxford
Marchini S, Macdonald DW (2012) Predicting ranchers' intention to kill jaguars: case studies in Amazonia and Pantanal. Biol Conserv 147:213–221
Marker L (2002) Aspects of cheetah (*Acinonyx jubatus*) biology, ecology and conservation strategies on Namibian farmlands. Department of Zoology, University of Oxford, Oxford
Marker L, Dickman A (2004) Human aspects of cheetah conservation: lessons learned from Namibian farmlands. Hum Dimens Wildl 9:1–9
Martin J, Runge MC, Nichols JD, Lubow BC, Kendall WL (2009) Structured decision making as a conceptual framework to identify thresholds for conservation and management. Ecol Appl 19:1079–1090
Mazzolli M, Graipel ME, Dunstone N (2002) Mountain lion depredation in southern Brazil. Biol Conserv 105:43–51
Messmer TA (2000) The emergence of human–wildlife conflict management: turning challenges into opportunities. Int Biodeterior Biodegrad 45:97–102
Messmer TA (2009) Human–wildlife conflicts: emerging challenges and opportunities. Hum Wildl Confl 3(1):10–17
Minnis D, Peyton RB (1995) Cultural carrying capacity: modeling a notion. In: Proceedings of the Urban Deer Symposium, St. Louis
Mishra C (1997) Livestock depredation by large carnivores in the Indian Trans-Himalaya: conflict perceptions and conservation prospects. Environ Conserv 24:338–343
Moreira JR, Ferraz KMPMB, Herrera EA, Macdonald DW (2012) Capybara: biology, use and conservation of an exceptional neotropical species. Springer, Berlin
Naughton-Treves L, Treves A (2005) Socio-ecological factors shaping local support for wildlife crop-raiding by elephants and other wildlife in Africa. In: Woodroffe R, Thirgood S, Rabinowitz A (eds) People and wildlife: conflict or coexistence? Cambridge University Press, Cambridge, pp 252–277
Nowell K, Jackson P (1996) Wild cats: status survey and conservation action plan. Burlington Press, Cambridge
Nyhus P, Osofsky S, Ferraro PJ, Madden F, Fischer H (2005) Bearing the costs of human–wildlife conflict: the challenges of compensation schemes. In: Woodroffe R, Thirgood S, Rabinowitz A (eds) People and wildlife: conflict or coexistence? Cambridge University Press, Cambridge, pp 107–121
Perez E, Pacheco LF (2006) Damage by large mammals to subsistence crops within a protected area in a montane forest of Bolivia. Crop Protect 25:933–939
Rabinowitz A (2005) Jaguars and livestock: living with the world's third largest car. In: Woodroffe R, Thirgood S, Rabinowitz A (eds) People and wildlife: conflict or coexistence? Cambridge University Press, Cambridge, pp 278–285

Rasmussen GSA (1999) Livestock predation by the painted hunting dog Lycaon pictus in a cattle ranching region of Zimbabwe: a case study. Biol Conserv 88:133–139
Reading RP, McCain L, Clark TW, Miller BJ (2005) Understanding and resolving the black-tailed prairie dog conservation challenge. In: Woodroffe R, Thirgood S, Rabinowitz A (eds) People and wildlife: conflict or coexistence? Cambridge University Press, Cambridge, pp 209–223
Redpath SM, Young J, Evely A, Adams WM, Sutherland WJ, Whitehouse A, Amar A, Lambert RA, Linnell JDC, Watt A, Gutierrez RJ (2013) Understanding and managing conservation conflicts. Trends Ecol Evol 28:100–109
Riley SJ, Decker DJ, Carpenter LH, Organ JF, Siemer WF, Mattfeld GF, Parsons G (2002) The essence of wildlife management. Wildl Soc Bull 30:585–593
Roskaft E, Handel B, Bjerke T, Kaltenborn BP (2007) Human attitudes towards large carnivores in Norway. Wildl Biol 13:172–185
Sanyal P (1987) Managing the man-eaters in the Sundarbans tiger reserve of India: a case study. In: Tilson RL, Seal US (eds) Tigers of the world: the biology, biopolitics, management and conservation of an endangered species. Noyes Publications, Park Ridge, pp 427–434
Sinclair ARE, Fryxell JM, Caughley G (2006) Wildlife ecology, conservation, and management. Blackwell Science, Cambridge
Soto-Shoender JR, Giuliano WM (2011) Predation on livestock by large carnivores in the tropical lowlands of Guatemala. Oryx 45:561–568
Swenson JE, Andren H (2005) A tale of two countries: large carnivore depredation and compensation schemes in Sweden and Norway. In: Woodroffe R, Thirgood S, Rabinowitz A (eds) People and wildlife: conflict or coexistence? Cambridge University Press, Cambridge, pp 323–339
Thirgood S, Redpath S (2005) Hen harrier and red grouse: the ecology of a conflict. In: Woodroffe R, Thirgood S, Rabinowitz A (eds) People and wildlife: conflict or coexistence? Cambridge University Press, Cambridge, pp 192–208
Thirgood S, Woodroffe R, Rabinowitz A (2005) The impact of human–wildlife conflict on human lives and livelihoods. In: Woodroffe R, Thirgood S, Rabinowitz A (eds) People and wildlife: conflict or coexistence? Cambridge University Press, Cambridge, pp 13–26
Treves A, Naughton-Treves L (2005) Evaluating lethal control in the management of human–wildlife conflict. In: Woodroffe R, Thirgood S, Rabinowitz A (eds) People and wildlife: conflict or coexistence? Cambridge University Press, Cambridge, pp 86–106
Treves A, Wallace RB, White S (2009) Participatory planning of interventions to mitigate human–wildlife conflicts. Conserv Biol 23:1577–1587
Treves A, Naughton-Treves L, Harper E, Mladenoff D, Rose R, Sickley T, Wydeven A (2004) Predicting human–carnivore conflict: a spatial model derived from 25 years of wolf predation on livestock. Conserv Biol 18:114–125
Tversky A, Kahneman D (1974) Judgement under uncertainty: heuristics and biases. Science 185:1124–1130
Veeramani A, Jayson EA, Easa PS (1996) Man-wildlife conflict: cattle lifting and human casualties in Kerala. Indian Forester 122:897–902
Wang SW, Macdonald DW (2006) Livestock predation by carnivores in Jigme Singye Wangchuck National Park, Bhutan. Biol Conserv 129:558–565
Wilson MA (1997) The wolf in Yellowstone: science, symbol, or politics: deconstructing the conflict between environmentalism and wise use. Soc Nat Resour 10:453–468
Woodroffe R, Ginsberg J, Macdonald DW (1997) The African wild dog: status survey and conservation action plan. IUCN/SSC Canid Specialist Group, Gland
WHO (2010) World health statistics 2010. World Health Organization, Geneva
Yirga G, Bauer H, Worasi Y, Asmelash S (2011) Farmers' perception of leopard (*Panthera pardus*) conservation in a human dominated landscape in northern Ethiopian highlands. Int J Biodiver Conserv 3(5):160–166

Zeller K (2007) Jaguars in the new millennium data set update: the state of the jaguar in 2006. Wildlife Conservation Society, New York

Zimmermann A, Baker N, Inskip C, Linnell JDC, Marchini S, Odden J, Rasmussen G, Treves A (2010) Contemporary views of human–carnivore conflicts on wild rangelands. In: du Toit J, Kock R, Deutsch JC (eds) Wild rangelands: conserving wildlife while maintaining livestock in semi-arid ecosystems. Wiley-Blackwell, Oxford, pp 129–151

Chapter 14
BIOTA/FAPESP: The Biodiversity Virtual Institute—Translating Research on Biodiversity and Ecosystem Services into Policies in a Megadiverse Country

Carlos A. Joly

Abstract In ten years, with an annual budget of approximately 3 million USD, the BIOTA/FAPESP Program supported 90 major research projects—which trained successfully 172 undergraduates, 169 M.Sc., 108 Ph.D. students, as well as 79 post-docs. Produced and stored information about approximately 12,000 species and managed to link and make available data from 35 major biological collections of the State of São Paulo. This effort is summarized in more than 600 articles published, in 180 scientific journals from which 110 are indexed by the Institute for Scientific Information (ISI) database. Among the indexed periodicals, Nature and Science have the highest impact factor, and the median value among all indexed periodicals that authors of the Biota Program have published was equal to 1.191, significantly higher than the average for the area in Brazil. Furthermore, the Program published, so far, 16 books and 2 Atlas. These results were used by the State of São Paulo Government to improve public policies of biodiversity conservation and restoration.

14.1 A Regional Overview of Biodiversity

The neotropical region that stretches from southernmost North America through to southernmost South America, thus encompassing most of the Latin American countries, is one of the most diverse biogeographic regions on Earth (Muñoz and Mondini 2008).

Paleogeographic evolution of the neotropical region over more than 100 Mya fostered an increasing compartmentalization and resulted in a marked increase in biome and habitat diversity throughout the cretaceous, tertiary, and quaternary.

C. A. Joly (✉)
Plant Biology Department, Biology Institute, State University of Campinas, Campinas, SP, PO Box 6109, 13083-970, Brazil
e-mail: cjoly@unicamp.br

L. M. Verdade et al. (eds.), *Applied Ecology and Human Dimensions in Biological Conservation*, DOI: 10.1007/978-3-642-54751-5_14,

The arrival of humans, 14,500 BP, was followed by intensive cultural diversification and mostly non-intensive land use. Up until pre-Colombian times, the physiographic evolution of the region together with the outstanding cultural diversification of the Amerindians, reflected in hundreds of languages, generally favored the accumulation of biodiversity and related cultural knowledge. A reverse trend was set into motion in post-Colombian time, culminating in today's large-scale agriculture, plantation forestry, and increasing urbanization. In 2006, the UN Population Division projected that in 2050 Latin America, urban population will exceed the entire population living in the region today (Arroyo et al. 2009). On the other hand, the surviving Amerindians are assembled into 400 groups, representing 34 language families and two special language groups (Montenegro and Stephens 2006) and represent a mere 1.6 % of the world's population, and 7 % of the total population of Latin America today.

The neotropical region monopolizes the planet's biodiversity due to: diversity of biogeographical divisions, diversity of ecosystems, diversity of species, diversity of life forms and functional groups, concentration of endemic organisms, and agro-biodiversity associated with cultural diversity.

Some highlights are: six countries of the neotropical region fall into the mega-diverse league; 32 % of global biodiversity in vascular plants, summing to an estimated 95,000, for a land area constituting 9.6 % of total land area worldwide; in South America: 33 % of global biodiversity in birds, 32 % of anurans, 25 % of mammals, and 20 % of reptiles; two Vavilovian centers of origin of agriculture and plant domestication; seven of the 25 biodiversity hot spots for conservation priority; a recently discovered hot spot for bryophytes at the extreme southern end of South America; 22 % of global Frontier Forest. Brazil, the largest country in the region, has an estimated 170–210 thousand described species considering all taxonomic groups, but is believed to have around 1.8 million in total, taking into account microorganisms and fungi (Lewinsohn and Prado 2005). If we consider only vascular plants, the country holds 13 % of the world's flora.

Main threats to biodiversity of the region are deforestation, fire, overexploitation, the introduction of exotic species, climate change, and pollution. It is particular worrying that: South America suffered the greatest ever-net forest reduction over the years 2000–2005; the Brazilian Cerrado is now disappearing at more than twice the rate as the Amazon rainforest; and rates of deforestation in other megadiverse countries like Mexico are still very high. Neotropical terrestrial, freshwater, and marine habitat have already received large numbers of exotic species, spanning the taxonomic hierarchy, but our knowledge regarding specific impacts on biodiversity is woefully incomplete.

Climate warming should lead to easier poleward migration of species in the northern extreme than in the southern part of the neotropical region, as a result of the fact that the amount of land increases with an increase in latitude north of the tropics, while in the South America south of the equator, the opposite is true. Results of the first Biodiversity modeling and climate change studies on the impacts of climate change suggest certain losses of biodiversity, along with complex feedbacks between drivers such as deforestation and climate change, leading to an exacerbation of global warming (Colombo and Joly 2010). However, experimental studies are still few, and overall, biodiversity scientists in Latin America, particularly ecologists, have been slow to rise to the challenge of tackling, large-scale, complex problems through networking and data sharing (Arroyo et al. 2009).

As been pointed out by a recent review of ICSU-LAC (Arroyo et al. 2009), huge asymmetries with respect to basic knowledge and/or its accessibility characterize marine and freshwater versus terrestrial habitats. A serious problem in general concerns the lack of georeferenced biodiversity data and the willingness of institutions, with some notable exceptions (e.g., CONABIO, INBio, BIOTA/FAPESP), to make data available online. The study of ecosystems services is hindered by the lack of data on carbon sequestration; nevertheless, economic valuations of some ecosystem services are beginning to appear, and ecotourism and its variants are well developed in the region. Climate change research at an ecosystem level is hindered by the lack of long-term data sets and the compilation of regional data sets, although there are some notable exceptions.

Close to 8,500 plants and animals in the neotropical region are considered to have conservation problems by IUCN standards, but this number is concluded to grossly underestimate the real situation. The most threatened groups are amphibians (32 % of total) and fishes (24 %); however, the vast majority of species cataloged as endangered (67 %) are plants. Although 21 % of the neotropical region land area is protected—the highest percentage contribution for all developing regions of the world, and higher than in the developed countries—distribution modeling and GAP analysis reveal that the present configuration of protected areas is not always optimally located to protect the region's biodiversity. Moreover, there are huge imbalances comparing the protection of wet forest habitats versus dry forest and scrubland habitats, represented, for instance, by the Cerrado, and the protection of terrestrial habitats versus marine habitats (Arroyo et al. 2009).

The vast and biologically rich neotropical region presents an outstanding opportunity to develop biodiversity science in many different dimensions. An overview of institutional arrangements and resources for biodiversity research shows that within the neotropical region, there are many institutions devoted, at least in part, to biodiversity science, among which are found several novel institutions of international standard fully devoted to biodiversity research (Arroyo et al. 2009).

14.2 The BIOTA/FAPESP Program

Within this scenario, in April 1996, the scientific community, working within the large umbrella that encompasses characterization, conservation, and sustainable use of the biodiversity, started to work on the profile of a research program aiming at solving these problems. Three years later, in March 1999, the **State of São Paulo Research Foundation/FAPESP** (http://www.fapesp.br) launched the **BIOTA/FAPESP Program: The Virtual Institute of Biodiversity** (http://www.biota.org.br).

The State of São Paulo, located in the Southeastern region of Brazil, is the most industrialized State of the country and has a population of over 40 million people. It currently presents urban and industrial development rates comparable to those of Western European countries, such as Spain, Italy, UK, France, and Germany. São Paulo has a population of 41, 541, 191 inhabitants, around 22 % of Brazil's population, a demographic density of 135 persons per km^2, three big metropolitan areas, and the most complex urban network of Latin America. São Paulo's gross state product (GSP) is ≈ US$ 450 billions with a per capita income of ≈ US$ 10,000/year. Currently, the State has 645 municipalities and the largest transport system of Latin America, with links between highways, railways, airports, and waterways, interconnecting all municipalities and cities with other Brazilian States, as well as with the majority of the Mercosul countries. The State accounts for 33.4 % of Brazil's gross national product (GNP) and 42 % of the total Brazilian exports, 11 % of non manufactured products, and 42 % of industrialized goods. Approximately 92 % of São Paulo exports concern industrialized goods—including airplanes (EMBRAER), cars, trucks, and buses. The State of São Paulo also contributes with significant part of the Brazilian chemical industry, with net sales of US$ 103.5 billion in 2008, a new record for the country, becoming one of the 10 largest in the world. It is also Brazilian's biggest sugarcane producer (270 million/tons/year), corresponding to 70 % of Brazilian's exports (US$ 5.65 billions in 2007) and is expected to increase another 50 % in the next 5 years.

Due to its geographical position, the transition between the tropical and subtropical region, its relief, with a large mountain range—the Serra do Mar—separating the always wet Coastal Plain (rainfall up to 2,200 mm) from the more seasonal Western Plateau (rainfall up to 1,400 mm), average temperatures varying from 18 to 28 °C, and its complex drainage network (with four major rivers—Tiete, Paraíba do Sul, Ribeira de Iguape, and Paranapanema), the biodiversity of the State of São Paulo is among the highest in the country.

Like Europe in between 1,500 and 1,800 (Kaplan et al. 2009), USA between 1,800 and 1,900 (Tchir and Johnson 2004), São Paulo State lost most of its native Atlantic Forest, due to the expansion of coffee plantations, and most of its native savannah (Cerrado) was lost due to sugar cane expansion during the first oil crises in early 1970s. As a result, the two major biomes of the State, Atlantic Forest and Cerrado (Savannah), have been reduced to 12 and 2 %, respectively. With the exception of the coastal mountains (Serra do Mar), which was too steep for coffee plantation, native vegetation in São Paulo State is highly fragmented needing extensive restoration to reconnect fragments and improve their biodiversity conservation capacity. There are only 230,000 ha of native Cerrado left, and these are pulverized in over 8,000 fragments, more than 4,000 of them with less than 10 ha, and only 47 with an area greater than 400 ha (Kronka et al. 1998; Cavalcanti and Joly 2002).

The relevance of biodiversity conservation in these two biomes, Atlantic Forest and Cerrado, has been internationally recognized with their inclusion in the list of "hot spots" (Myers et al. 2000). Therefore, it is not surprising that the biodiversity numbers of the State are extremely high, around 7,000 species of higher plants

(Wanderley et al. 2011) more than 2,000 of vertebrates (Oyakawa and Menezes 2011; Menezes 2011; Rossa-Feres et al. 2011; Zaher et al. 2011; Silveira and Uezu 2011; de Vivo et al. 2011), more than 500,000 of invertebrates and the number of microorganisms can only be speculated. At least 30 % of these species are endemic, what makes even more urgent the development of tools to, simultaneously, increase our knowledge, establish sound conservation policies, and learn to use this natural treasure in a sustainable way.

As in other part of the world, one of the major problems to improve public policies on biodiversity conservation and sustainable use is the fact that the information already available is generally fragmented, disperse, of difficult access and, consequently, underused. Besides, as a consequence of the lack of an updated cartographic base, the location of sampling sites, a key information, is usually inaccurate. The greatest challenge is to systematize sampling, using GPS to locate the sampling site/area, develop an integrated databank for storing this information, and to produce accurate and reliable maps of geographical occurrence and distribution of species.

In April 1996, the scientific community, working within the large umbrella that encompasses characterization, conservation, and sustainable use of the biodiversity, started to work on the profile of a research program aiming at solving these problems. Three years later, in March 1999, the State of São Paulo Research Foundation/FAPESP (http://www.fapesp.br), a non-political public foundation funded by taxpayers in the State of São Paulo, charged with enabling scientific research and technological development in all areas of knowledge, launched the research program on characterization, conservation, and sustainable use of the biodiversity of the State of São Paulo, named BIOTA/FAPESP—The Biodiversity Virtual Institute (http://www.biota.org.br).

14.2.1 The Creation of the BIOTA/FAPESP Program

The first problem to be tackled was the development of tools and means to increase connections among researchers and research institutions working with biodiversity (Speglich and Joly 2003). Therefore, a homepage (http://www.biota.org.br) and a discussion list were the first steps. Through the discussion list, we had a long and very fruitful discussion about the importance of making information on biodiversity knowledge available to public access via Internet.

The most important issue from this discussion was concerning copyrights of, for example, a list of birds, or fishes or plants of São Paulo State published only in the Internet. Once this was solved, by tagging to the "online" publication a metadata label with the copyright information, we started publishing the available species lists for the State.

These lists were a starting point for a thorough inventory of the available knowledge about our native biodiversity. Taking into account that species from São Paulo State (mainly of vertebrates and higher plants) have been recorded, collected and described since early 1800s by European expeditions, we decided that it was important to evaluate the existing knowledge about different taxonomic groups, ranging from virus to mammals and angiosperms, as well as the list of personnel and institutions working with each taxonomic group, and the State *ex situ* and *in situ* infrastructure for their conservation. At that stage, there were approximately 70 researchers involved.

In order to consolidate these inventories and discuss how to start a cooperative effort to study the biodiversity of the State, in July 1997, we organized a workshop, with over 100 participants from many research areas and institutions. The quality of the documents prepared for that meeting encouraged us to publish them in a series of 7 volumes named *Biodiversity of the State of São Paulo: a synthesis of knowledge at the end of the twentieth century (Biodiversidade do Estado de São Paulo: síntese do conhecimento ao final do século XX*) and to make them fully available through the Internet (http://www.biota.org.br/publi). During that meeting, we defined as long-term common objective for all the BIOTA/FAPESP research projects, the study of the biodiversity (using the broadest definition of biodiversity as stated in the CBD) of the State of São Paulo aiming:

(a) To invent and characterize the biodiversity of the State of São Paulo and to define the mechanisms for its conservation and sustainable use;
(b) To understand the processes that generate and maintain biodiversity, as well as those that can result in its deleterious reduction;
(c) To standardize sampling, making the use of GPS mandatory;
(d) To make information relevant to conservation and sustainable use of biodiversity available to decision makers;
(e) To ensure fast and free public access to this information;
(f) To improve teaching standards on subjects related to conservation and sustainable use of biodiversity.

The research projects linked to the Program were conducted in order to increase the academic knowledge about the States' biodiversity and to, simultaneously, produce data potentially useful for improving State policies on biodiversity conservation and sustainable use.

All major public universities (USP, UNICAMP, UNESP, UFSCar, UNIFESP), some private universities (such as PUC, UNAERP, UNITAU, UMC, and UNISANTOS), research institutes (such as the Instituto de Botânica, Instituto Florestal, Instituto Geológico, INPE), EMBRAPA centers, and NGOs (such as Instituto Socioambiental, Fundação SOS Mata Atlântica, Conservation International and Reference Center on Environmental Information/CRIA) took part in the first 10 years of the Program. Considering just researchers linked to those institutions within the State of São Paulo, the BIOTA/FAPESP community brings together approximately 500 Ph.D.s, plus 400 graduate students. In addition, there are 100 collaborators from other Brazilian States and approximately 80 from abroad.

An important feature of the BIOTA/FAPESP Program is that the researchers involved are conducting their research on areas of their specific training and skill, but all of them have added common goals to their projects. Furthermore, they are using a set of common tools that have been developed for integrating data within the BIOTA/FAPESP Program.

The BIOTA/FAPESP Program was inspired by the Convention on Biological Diversity (CBD) and employs its definition of biodiversity from terrestrial and aquatic (including marine and freshwater) ecosystems. The scope of research developed under the program ranges from DNA barcoding to landscape ecology, including taxonomy, phylogeny, phylogeography, and the human dimensions of biodiversity conservation and sustainable use. The major aims of BIOTA/FAPESP are as follows: (a) to invent and characterize the biodiversity of the State of São Paulo while defining the mechanisms for its conservation and sustainable use; (b) to understand the process that generate and maintain biodiversity, as well as those that result in biodiversity loss; (c) to organize and make available to policy makers and to society in general, biological information relevant for the establishment of priorities for biodiversity conservation and sustainable use; (d) to have all information freely available through the Internet; (e) to improve teaching standards and public awareness on subjects related to biodiversity conservation and sustainable use.

During the first 10 years term (1999–2008), the Program has averaged an US$ 3,000,000 yearly budget from FAPESP, supported 94 major research projects, trained 169 M.Sc., 108 Ph.D. students, and 79 post-docs, described more than 1,800 new species, acquired and archived information on over 12,000 species, and linked and made available data from 35 major biological collections. This effort has produced more than 750 articles published in international scientific journals, 18 books, and 2 atlases. Over 1,200 researchers and students are currently engaged in BIOTA projects. While most are from São Paulo, there are at least 150 collaborators from other Brazilian States and 100 from abroad. BIOTA launched a new electronic journal in 2001, *Biota Neotropica* (http://www.biotaneotropica.org.br), which is currently indexed by ISI, and a new venture in 2002 called BIO*prospec*TA (http://www.bioprospecta.org.br) that has already submitted three new drug patents.

14.2.2 The Environmental Information System/SinBiota[1]

The establishment of a standard record form to register sampling data also enhanced connectivity among projects. All research teams discussed this record form during almost a year, before reaching final agreement on the mandatory fields. In the end, **nine mandatory fields** were established: **sampling author; date; locality, including the geographical coordinates obtained by GPS;**

[1] http://sinbiota.biota.org.br/sinbiota

municipality; watershed; taxa[2]; sampling method; ecosystem. Besides these nine mandatory items, there are more than forty supplementary fields that can be used to give more detailed information, if required, regarding the specific taxonomic group or research.

As the result of a collective effort, these tools (standard sampling form and standard form for species lists) have proved to be suitable to all new research projects and are strongholds of the BIOTA/FAPESP Program. They were also essential to the construction of a databank for registering all samples collected by researchers participating in the Program.

The first Environmental Information System (http://sinbiota1.biota.org.br/) was developed by the Reference Center on Environmental Information/CRIA (http://www.cria.org.br) in collaboration with the Instituto de Computação (http://www.ic.unicamp.br) of the State University of Campinas/UNICAMP (http://www.unicamp.br), open source computational languages, and software such as Linux and PostgreSQL. Another important feature of *SinBiota* is its full interoperability with other Brazilian (such as SpeciesLink http://splink.cria.org.br/) or international (such as GBIF http://www.gbif.org/) initiatives.

Currently, the BIOTA/FAPESP Program is testing a new version for its Environmental Information System, the *SinBiota 2.0* (http://sinbiota.biota.org.br/sinbiota), which is hosted by the node of the Brazilian High-Performance Processing Network based at the State University of Campinas (CENAPAD/UNICAMP—http://www.cenapad.unicamp.br/).

Along with the development of the databank and its interface with Internet, a digital map of the State of São Paulo, in a 1:50,000 scale, was produced in collaboration with the Instituto Florestal (http://www.iflorestsp.br/) and UNICAMP (Instituto de Geociências http://www.ige.unicamp.br; Faculdade de Engenharia Agricola http://www.agr.unicamp.br and CEPAGRI http://www.cpa.unicamp.br). The map has detailed information about: urban areas; roads; county boundaries; rivers; areas covered by *Eucalyptus* spp. and *Pinnus* spp. wood exploitation forests; conservation units; and remnants of native vegetation. These remnants of native vegetation are classified in 37 categories, according to the official classification of Brazilian vegetation (Veloso et al. 1991).

The digital atlas is an assemblage of the 416 cartographic charts from the 1972 IBGE (Instituto Brasileiro de Geografia e Estatística http://www.ibge.gov.br) map of São Paulo State, updated with Landsat 5 or 7 satellite images from 2001/2003.

The geographic coordinates, one of the mandatory fields from the standard sampling form, connect the databank and the digital map, allowing, in this stage, a display "on the fly" of the spatial distribution of occurrence sites of species registered in the *SinBiota*. The system also allows zooms, besides the connection with the standard sampling form related to the sites plotted on the map, and the visualization of all the registered information concerning that species.

[2] A standard pattern of species lists was established for each recognized taxonomic group. Consequently, attached to the sampling site record the researcher send the associate list of taxon or taxa collected in that specific locality.

SinBiota 2.0 Atlas (http://sinbiota.biota.org.br/Sinbiota/Map/) gives also the possibility of using Bing Maps as the cartographic base to plot sampling spots and/or species distribution.

14.2.3 BIOTA Symposia and Evaluation Meetings

In spite of these electronic means of connecting research projects, researcher meetings are of paramount importance. We organize every 2 years a symposium, focused mainly in the undergraduate (B.Sc.) and graduate students (M.Sc. and Ph.D), as well as post-docs of the ongoing research projects of the Program.

Immediately after the symposium, usually with a one day of overlapping, we have the so-called evaluation meeting. This meeting is mainly for principal investigators (PIs) and a selected panel of international of 4–6 experts, invited by the Scientific Directory of FAPESP to evaluate the BIOTA/FAPESP Program. So, besides the individual evaluation of the progress of each research project, carried out on an ad hoc basis by FAPESP, the progress and evolution of the whole Program is evaluated every 2 years. The reports produced by this panel of international experts (available at http://www.biota.org.br/info/sac/) is one of the main tools of the Program′s Steering Committee to evaluate the need of new approaches and/or integration tools, in a continuous process of improvement of the Program.

14.2.4 Biota Neotropica [3]

In 2001, the Program launched the online only journal BIOTA NEOTROPICA, to publish results of original research, associated or not to the program, concerned with characterization, conservation, and sustainable use of biodiversity within the neotropical region.

Since its first number, BIOTA NEOTROPICA has been guided toward international standards, using a double blind ad hoc referee system, and increasing gradually the compulsory use of English. Therefore, currently it is indexed by Thompson's ISI, and became a top reference among Latin American biodiversity journals.

14.2.5 BIOprospecTA [4]

Last, but not least, in 2002, the Program launched its network of researchers and laboratories with the objective of establishing the basis for the sustainable use of our biodiversity.

---

[3] http://www.biotaneotropica.org.br

[4] http://www.bioprospecta.org.br/

The State of São Paulo has several research groups working on all areas which are important for a successful bioprospection program, with remarkable experience and proved competences. The Biota Program has brought together a large group of researchers involved in the taxonomical knowledge of our biodiversity. We have many groups working in isolation/purification and structure elucidation of natural metabolites; others with large experience in pharmacology with established in vitro and in vivo bioassays; others with excellent facilities and expertise for the rapid identification and characterization of compounds in crude extracts (e.g., NMR, crystallography, mass spectrometry, etc.). Nevertheless, there is a strong need for a better integration of these competences within the common goal of achieving the desired applications for our biodiversity.

Among the main objectives of the BIO*prospec*TA network are: standardized collection of biological samples (plants, microorganisms, marine species, insects, etc.) and preprocessing of raw materials for the subsequent preparation of extracts; creation of a bank of extracts and pure compounds from plants, microorganisms, marine organisms, and other natural sources, with the required automation and data management facilities; characterization of promising extracts/compounds (NMR, Crystallography, LC/GC-MS, etc.); screening of extracts with existing in vitro and in vivo bioassays; development of new bioassays, particularly those adequate for high-throughput screening using small sample volumes; pharmacology and toxicology of promising bioactive extracts or compounds; synthesis of bioactive natural products and their derivatives; in partnership with the productive sector medicinal chemistry and drug design applied to the development of promising compounds; last but not least, development of a database structure for the data processing of the program.

During the last 5 year, BIO*prospec*TA supported 16 projects, which published 180 papers and deposited four patents. Cosmetic and pharmaceutical companies already showed interest in a partnership to screen BIO*prospec*TA bank of extracts for specific targets.

14.2.6 Improving Public Policies of Biodiversity Conservation and Restoration

During 2006 and 2007, the BIOTA/FAPESP researchers, in collaboration with the State of São Paulo Secretary for Environment/SMA and Conservation International, made an extraordinary effort to synthesize its databank in a set of eight maps of biodiversity conservation and restoration priority areas in the State of São Paulo (Joly et al. 2010). Detailed biological and of landscape metrics information every single area indicated in these maps have been synthesized in the book **Guidelines for biodiversity conservation and restoration in the State of São Paulo** (Rodrigues et al. 2008).

These maps and the book have just been adopted by the government of the State of São Paulo as the legal framework for biodiversity conservation policies in the State. There are now more than 20 legal instruments that mention the

BIOTA/FAPESP Research Program in their introductory justification. One of the most striking implementations of BIOTA/FAPESP recommendations is a joint resolution of the State secretaries of the environment and of agriculture to establish an agro-ecological zoning ordinance that prohibits sugarcane expansion to areas that are priorities for biodiversity conservation and restoration (Joly et al. 2010). It is a rare example of how a large and well-planned research effort can be used to set environmental policies of an industrialized State such as São Paulo.

14.2.7 Publications and Human Resources

In ten years, with an annual budget of approximately 3 millions USD, the BIOTA/FAPESP Program supported 90 major research projects—which trained successfully 172 undergraduate, 169 M.Sc., 108 Ph.D. students, as well as 79 post-docs. Produced and stored information about approximately 12,000 species and managed to link and make available data from 35 major biological collections of the State of São Paulo. This effort is summarized in more than 750 articles published, in 180 scientific journals from which 110 are indexed by the Institute for Scientific Information (ISI) database. Among the indexed periodicals, Nature and Science have the highest impact factor, and the median value among all indexed periodicals that authors of the Biota Program have published was equal to 1.191, significantly higher than the average for the area in Brazil. Furthermore, the Program published, so far, 16 books and 2 Atlas.

14.2.8 Internet 2

Finally, it is important to register that "when establishing an agreement with the National Science Foundation (NSF) in order to enable the access of the whole research system of the State to the Internet 2 network of the USA and the rest of the world, FAPESP presented the cooperation between BIOTA and the Species Analyst project of the University of Kansas as an example of interaction that would demand such a communication facility. This is another product of the BIOTA Program with immediate benefits to the whole scientific community of São Paulo" (Perez 2002).

14.2.9 Planning the Next 10 Years

Revisiting the broad objectives of the BIOTA/FAPESP Program, we realize that some of them are long lasting and still prevail in similar initiatives around the world, such as DIVERSITAS (http://www.diversitas-international.org/) whose mission is: (a) promote an integrative biodiversity science, linking biological, ecological, and social disciplines in an effort to produce socially relevant new knowledge and (b) provide the scientific basis for the conservation and sustainable use of biodiversity. Nevertheless, after 10 years, we see the need to modernize methodologies and techniques as well as to bring new scientific challenges to broader the community of scientist potentially interested in joining the BIOTA/FAPESP Program and to keep producing high standard science. One of the biggest challenges of this new phase is to give the BIOTA/FAPESP Program a position in the international arena that matches the quality of the science we produce.

In June 2009, during two days (3rd and 4th of June—**Workshop BIOTA + 10: setting agenda and priorities for 2020**), more than 300 scientists and students associated to research projects within the theme biodiversity characterization, conservation restoration, and sustainable use, already linked or not to the BIOTA/FAPESP Program, discussed priorities and an agenda for the next 10 years of the Program.

As a result of this discussion, it was decided that the following objectives of the BIOTA/FAPESP Program will be further pursued in the next decade.

- To invent and characterize the biodiversity of the State of São Paulo, by defining the mechanisms for its conservation and sustainable use;
- To understand the processes that generate and maintain biodiversity, as well as those that can result in its deleterious reduction;
- To produce estimates about biodiversity loss in different spatial and timescales;
- To evaluate the effectiveness of conservation initiatives within the State of São Paulo, identifying priority areas and components for conservation;
- To increase the ability of the State of São Paulo and public and private organizations in managing, monitoring and using biodiversity in a sustainable way.

Furthermore, the following **(twelve) points** have been thoroughly discussed and elected as top priorities for the next 10 years.

1. **Including native biodiversity restoration as one main objective of the BIOTA/FAPESP Program.**
2. **Development and implementation of a new information system for the BIOTA/FAPESP Program.**
3. **Biodiversity inventories and DNA barcoding and phylogeography**.
4. **Increase coastal and marine biodiversity studies.**
5. **Invasive species and GMOs.**
6. **Landscape ecology and ecosystem functioning and services.**
7. **Applied ecology and human dimensions in biological conservation.**
8. **Biodiversity modeling and climate change impacts in biodiversity.**
9. **Short-, medium-, and long-term plans for the BIO*prospec*TA network.**
10. **Education and public outreach.**
11. **Short-, medium-, and long-term plans for BIOTA NEOTROPICA** (www.biotaneotropica.org.br).
12. **National and international partnerships.**

During the last 2 years, some of these points are already being accomplished. Good examples are: (a) the development of the new information system is progressing well, and its prototype is now being tested; (b) with a specific call of proposals, we managed to bring up to 15 research projects focused in coastal and marine biodiversity; (c) in 2012, the BIOTA/FAPESP Program made joint calls with national, CNPq/National Research Council, and international, National Science Foundation/NSF and Natural Environment Research Council/NERC.

Education and public outreach will be the main focus of the Program in 2013/14. In February 2013, the Program launched a series of conferences focused in the Brazilian Biomes having as main target High School teachers and students. All lectures and additional material are available in the Program's homepage for free download to be used by teachers and students. In September 2013, a large multisensory exposition, also focusing on Brazilian biomes, opened in São Paulo city and thereafter will be traveling to other cities.

The good performance of the BIOTA/FAPESP Program in all four fronts: advancing scientific knowledge on biodiversity and ecosystem services, capacity building, dialog with decision makers to improve public policies, and interlocution with the productive sector to license patents, lead FAPESP to renew its support to the Program until 2020. Currently, FAPESP supports 77 ongoing research projects and 114 scholarships (M.Sc., Ph.D. and Post-docs) within the BIOTA/FAPESP Program, and in average, FAPESP has quadruple the resources invested in the Program, investing over US$ 24 million in 2011 and 2012.

References

Arroyo MTK, Dirzo R, Joly CA, Castillas JC, Cejas F (2009) Biodiversity knowledge research scope and priority areas: an assessment for Latin America and the Caribbean. In: ICSU-LAC science for a better life: developing regional scientific programs in priority areas for Latin America and the Caribbean, vol 1. ICSU Regional Office for Latin America and the Caribbean, Rio de Janeiro and Mexico City

Cavalcanti RB, Joly CA (2002) Biodiversity and conservation priorities in the Cerrado region. In: Oliveira PS, Marquis RJ (eds) The Cerrados of Brazil: ecology and natural history of a neotropical savanna. Columbia University Press, New York, pp 351–367

Colombo AF, Joly CA (2010) Brazilian atlantic forest lato sensu: the most ancient Brazilian forest, and a biodiversity hotspot, is highly threatened by climate change. Braz J Biol 70(3): 697–708

De Vivo M, Carmignotto AP, Gregorin R, Hingst-Zaher E, Iack-Ximenes GE, Miretzki M, Percequillo AR, Rollo MM, Rossi RV, Taddei VA (2011) Checklist of mammals from São Paulo State, Brazil. Biota Neotropica 11(1a) (http://www.biotaneotropica.org.br/v11n1a/en/abstract?inventory+bn0071101a2011)

Joly CA, Rodrigues RR, Metzger JP, Haddad CFB, Verdade LM, Oliveira MC, Bolzani VS (2010) Biodiversity conservation, research, training and policy in São Paulo State. Science 328:1358–1359

Kaplan JO, Krumhardt KM, Zimmermann N (2009) The prehistoric and preindustrial deforestation of Europe. Quat Sc Rev 28:3016–3034

Kronka JNF, Nalon MA, Matsukuma CK, Pavão M, Guillaumon JR, Cavalli AC, Giannotti E, Ywane MSS, Lima L, Montes J, Cali IHD, Haack PG (1998) Áreas de domínio do Cerrado no Estado de São Paulo. São Paulo, Secretaria do Meio Ambiente

Lewinsohn TM, Prado PI (2005) How many species are there in Brazil. Conserv Biol 19:619–624
Menezes NA (2011) Checklist of marine fishes from São Paulo State, Brazil. Biota Neotropica 11(1a) (http://www.biotaneotropica.org.br/v11n1a/en/abstract?inventory+bn00311101a2011)
Montenegro R, Stephens C (2006) Indigenous health in Latin America and the Caribbean. Lancet 367:1859–1869
Muñoz AS, Mondini M (2008) Neotropical zooarchaeology and taphonomy. Quat Int 180:1–4
Myers N, Mittermeier RA, Mittermeier CG, Fonseca GAB, Kent J (2000) Biodiversity hotspots for conservation priorities. Nature 403:852–858
Oyakawa OT, Menezes NA (2011) Checklist of fresh water fishes from São Paulo State, Brazil. Biota Neotropica 11(1a) (http://www.biotaneotropica.org.br/v11n1a/en/abstract?inventory+bn00211101a2011)
Perez JF (2002) BIOTA-FAPESP: The different dimensions of success. Biota Neotropica 2(1) (http://www.biotaneotropica.org.br/v2n1/en/editorial)
Rodrigues RR, Joly CA, Brito MCW, Paese A, Metzger JO, Casatti L, Nalon MA, Menezes NA, Bolzani VS, Bononi VLR (2008) Diretrizes para a conservação e restauração da biodiversidade no estado de São Paulo. Programa BIOTA/FAPESP and FAPESP and Secretaria do Meio Ambiente
Rossa-Feres DC, Sawaya RJ, Faivovich J, Giovanelli JGR, Brasileiro CA, Schiesari L, Alexandrino J, Haddad CFB (2011) Amphibians of São Paulo State, Brazil: state-of art perspect. Biota Neotropica 11(1a) (http://www.biotaneotropica.org.br/v11n1a/en/abstract?inventory+bn00411101a2011)
Silveira LF, Uezu A (2011) Checklist of birds from São Paulo State, Brazil. Biota Neotropica 11(1a) (http://www.biotaneotropica.org.br/v11n1a/en/abstract?inventory+bn00611101a2011)
Speglich E, Joly CA (2003) The Brazilian Biodiversity Virtual Institute. In: Lemons J, Victor R, Schaffer D (eds) Conserving biodiversity in arid regions. Kluwer Academic Publishers, Boston, pp 381–386
Tchir TL, Johnson EA (2004) The history of deforestation in North America. In: UNESCO encyclopedia of life support systems (EOLSS). Paris
Veloso HP, Rangel Filho ALR, Lima JCA (1991) Classificação da Vegetação Brasileira, adaptada a um sistema universal. MEFP/IBGE/DRNEA, Rio de Janeiro
Wanderley MGL, Shepherd GJ, Martins SE, Estrada TEMD, Romanini RP, Koch I, Pirani JR, Melhem TS, Harley AMG, Kinoshita LS, Magenta MAG, Wagner HML, Barros F, Lohmann LG, Amaral MCE, Cordeiro I, Aragaki S, Bianchini RS, Esteves GL (2011) Checklist of Spermatophyta of the São Paulo State, Brazil. Biota Neotropica 11(1a) (http://www.biotaneotropica.org.br/v11n1a/en/abstract?inventory+bn01311101a2011)
Zaher H, Barbo FE, Martínez PS, Nogueira C, Rodrigues MT, Sawaya RJ (2011) Reptiles from São Paulo State: Current knowledge and perspectives. Biota Neotropica 11(1a) (http://www.biotaneotropica.org.br/v11n1a/en/abstract?inventory+bn00511101a2011)

Index

A

Abalone, 77
Abundance, 8–11, 13, 93, 117, 136, 137
  absolute, 8, 10, 11
  estimates, 121–123, 125, 126, 128, 129
  indexes, 137
  patterns, 4, 9, 10, 13
  relative, 8, 10, 11, 124
Acclimation, 53, 58–64, 67
Accuracy, 136, 150
Adaptation, 9, 11, 13, 53, 58–62, 64–67, 97
Adaptive Management (AM), 118, 120
Administration, 12, 13
Agriculture, 54, 55, 57, 59, 61–64, 66, 67, 92, 93, 98, 134–136, 138
  agricultural impacts, 98
  agricultural landscapes, 54, 55, 59, 63, 93, 95, 97, 98, 150, 151
  intensification of agricultural practices, 93, 95
  land sparing, 95
  multi-functionality of agricultural landscapes, 98
  subsistence agriculture, 134, 135
Agroecosystems, 94, 95, 97
  residents, 94
Agroforestry systems, 134
Amazonia, 23
Amphibians, 77
Animal movements, 159, 160, 165
Anthropogenic environments, 97
Applied ecology, 7, 12

B

Biodiversity, 5, 8, 9, 12, 74, 93, 95, 98, 136, 180
  conservation, 8, 12, 95, 97, 134
  loss, 4, 5, 9, 97
  modeling and climate change, 212, 222
  monitoring, 8–11, 13
Biomagnification, 168
Biomass, 7, 11, 12, 14, 182
Bioprospection program, 220
Biota Neotropica, 217, 219, 222
Birds, 82
  survey, 150
Black-footed ferrets, 76
Black swans, 178
Bootstrap, 152, 198, 212, 214, 221
Brazilian Forest Code, 95
Brazilian spotted fever, 193

C

Camera trapping, 136–138, 140
Canine distemper, 76
Captive breeding, 84
Capture-recapture, 127, 141
Carbon, 182
Carnivore, 106–112
  African lions, 108
  Attacks on humans, 196
  black and brown bears, 108
  black bears, 108
  bobcat, 108
  brown bears, 108
  coyotes, 109
  dogs, 77
  European badgers, 108
  feces, 110
  fur seals, 110
  Iberian lynx, 108
  jaguars, 108
  leopard cat, 108, 110

L. M. Verdade et al. (eds.), *Applied Ecology and Human Dimensions in Biological Conservation*, DOI: 10.1007/978-3-642-54751-5,

pumas, 108, 110
swift foxes, 108
wolverine, 109
wolves, 83
Carrying capacity
cultural, 203
Chemical processes, 160
Chytrid fungus, 77
Climate, 177
Climate change, 79, 95
Closed population models, 127
Collagen, 162
bones collagen, 167
Community, 8, 150
Compensation payments, 195
Complexity, 11
Complexity of processes, 10, 12, 13
Conservation, 95, 117, 118, 125, 129, 136
biological conservation, 4–7, 10, 12, 98
conservation biology, 97, 160
Contaminants, 167, 168
Control, 7, 8, 12
Count, 123
Counting calls, 136, 141
Crop-raiding, 191, 192
Culling, 84

D
Data, 180
Deforestation, 151
Deliberative phase, 121
Density, 128, 136
Design
experimental, 6, 7
sampling, 8, 9, 12
Detectability, 150, 180
Detection, 150, 152, 155, 156
probability, 123, 124
Diamondback terrapin, 80, 159–161
Dilution effect, 79
Direct observations, 125
Disease, 10, 73
Dispersal, 94
Distance sampling, 126
Distribution, 136, 182
patterns, 4, 9, 10
species, 9, 10, 13
Diversity
α-diversity, 98
β diversity, 98, 138
biological, 7, 10, 11
patterns, 9, 10, 12, 13
phylogenetic, 10
taxonomic, 10
Domestication, 93
domesticated species, 93
domestic species, 93
Double-loop learning, 121

E
Ecological processes, 97
Ecosystem, 184
Ecosystem services, 36–39, 46, 98
Ecotoxicology, 159, 160, 167, 168
Ethical dilemmas, 205
Eucalyptus, 151, 154
Evolution, 7, 12, 55, 58, 64, 97
evolutionary process, 95
groups, 10
lineages, 10
process, 7, 9, 10, 13
question, 4
rapid, 97
success, 5
Eucalyptus plantations, 55, 64
Evosystem services, 39, 98
Exotic species, 134
invasive, 93
Experimental design, 179
Extinction, 80, 93, 134
local, 94
rate, 93

F
Famine-feast, 97
Feces, 136, 137, 141
Fishing, 79
Food resources, 97
Food webs, 163, 165, 167, 74
Footprints, 136
Footprint surveys, 137
Forestry, 136, 182
Fractionation, 164
isotopic fractionation, 160, 161
Frequency of occurrence, 154

H
Habitat, 94
quality, 134
use, 136
Habitat destruction, 54, 56–58, 93, 134
large dam, 54

mining, 54
oil and gas extraction, 54
small hydroelectric power plants, 54
Heterogeneity
spatial, 6, 7, 9, 13, 97
temporal, 6, 7, 9, 13, 139
Historical ecology, 4, 19
Human dimensions, 9, 12, 13, 199
Human fatalities, 193
Humanism, 4, 5
Human-wildlife conflict, 190
Hunting, 79

I
Impacts, 199
Incidence-based species accumulation curves, 137
Indicator, 8, 14
landscape, 12
robust, 12
species, 82
Indices, 123, 124
Infrastructure, 180
Innovation, 8, 10, 12, 13
Interviews, 138, 140
Intraspecific competition, 162
Island fox, 84

L
Land use change (LUC), 9, 11, 12
Landscape, 4, 7, 11–13, 53, 134, 136
agricultural, 9
agro-forestry landscapes, 143
connectivity, 134
ecology, 4
fragmentation, 97, 134
matrix, 94, 95
matrix permeability, 134
matrix quality, 134
structure, 94
transformation, 19
Learning, 120

P
Pantanal, 192
Parasite, 10, 73
Partula turgida, 76
Passenger pigeon, 80
Pastures, 151
Pathogen/parasite-host, 11
Pathogens, 10
PD complementarity, 40, 45
PD-species relationship, 42
Persecution, 196
Pests, 190
control, 97
Pet trade, 84
Phylogenetic diversity
PD dissimilarity, 40
Point
counts, 150, 151, 155
Policy-making process, 10
Pollution, 78
Population, 4–7, 11–13
decline, 6, 7
declining population paradigm, 6
density, 10
growth, 8, 10
human, 9, 10
regulation, 75
size, 11
small population paradigm, 6
Population/community, 12
PPBio, 180
Precision, 136, 150
Predator-prey interactions, 11
Primates, 80
Publications, 184
Public policy, 95

R
Radioactive decay, 160
Radiotelemetry, 165
Radiotracking, 142
RAPELD, 177
Recurrent decisions, 120
Recurrent decision problems, 119
Remote sensing, 180
Reserva Ducke, 180

S
Sample units, 123
Sampling design, 139, 151
transect line, 150
Scales, 184
Scent stations, 136, 142
Sharks, 82
Spatially explicit capture-recapture models, 128
Species extinction, 54, 56
Species incidence curve, 11, 152

Species richness, 136, 138
Squirrels, 77
Stable isotope, 11, 13, 160, 165, 169
  abundance of stable isotopes, 169
  analysis, 169
  carbon and nitrogen stable isotopes, 162
  hydrogen stable isotope, 163, 165
  methodology, 162
  stable isotope analysis, 159–163, 166–169
  stable isotope ratio, 160, 164
  stable isotopes composition, 160, 162, 164–166
  stable isotope technique, 162, 165
  stable isotopes tool, 160
Stakeholder, 179, 202
Standardized, 179
State of São Paulo Research Foundation (FAPESP), 213
State of São Paulo, 213
State variables, 122
Stress, 78

T
Tasmanian devil, 84
TERN-ACEAS, 184
Time-of-detection models, 126
Tissues, 164–167, 169
  animal tissues, 161, 165, 167
  consumers tissues, 163
  fish tissues, 162
  inert tissues, 165
  metabolically active tissues, 165
  metabolically inert tissues, 165
Track plot, 136–138, 140
Tragedy of the commons, 95
Trematodes, 82
Trophic, 13
  level, 11, 13, 159, 160, 163, 164
  process, 11
  structure, 10–12, 95
Trout, 76
Turnover rates, 159, 160, 166, 167

V
Vaccination, 84
Viable populations, 95
Visualization, 152, 154, 155
Vocalization, 152, 154, 155

W
Whales, 81
Wicked problem, 204
Wildlife, 94, 136
  wildlife-friendly farming, 95
  monitoring, 137
  studies, 159–161, 166, 169
  surveys, 138

The manufacturer's authorised representative in the EU is Springer Nature Customer Service Centre GmbH, Europaplatz 3, 69115 Heidelberg, Germany. If you have any concerns regarding our products, please contact ProductSafety@springernature.com

Printed and bound by CPI Group (UK) Ltd, Croydon, CR0 4YY
22/09/2026
02221063-0012